Living with the shore

Series editors

Orrin H. Pilkey, Jr.
William J. Neal

The beaches are moving: the drowning of America's
shoreline, *new edition*

Wallace Kaufman and Orrin H. Pilkey, Jr.

From Currituck to Calabash: living with North Carolina's
barrier islands, *second edition*

Orrin H. Pilkey, Jr., William J. Neal, Orrin H. Pilkey, Sr., and
Stanley R. Riggs

Living with the Texas shore

Robert A. Morton, Orrin H. Pilkey, Jr., Orrin H. Pilkey, Sr., and
William J. Neal

Living with the South Carolina shore

William J. Neal, W. Carlyle Blakeney, Jr., Orrin H. Pilkey, Jr., and
Orrin H. Pilkey, Sr.

Living with the Louisiana shore

Joseph T. Kelley, Alice R. Kelley, Orrin H. Pilkey, Sr., and
Albert A. Clark

Living with Long Island's south shore

Larry R. McCormick
Orrin H. Pilkey, Jr.
William J. Neal
Orrin H. Pilkey, Sr.

Duke University Press   Durham, North Carolina   1984

Publication of this book was subsidized in part by a grant from the American Conservation Association.

Printed in the United States of America

Library of Congress Cataloging in Publication Data

McCormick, Larry, 1938–
  Living with Long Island's south shore.

  (Living with the shore)
  Includes bibliographical references and index.
  1. Shore protection—New York (State)—Long Island.
2. Coastal zone management—New York (State)—Long
Island.   I. Neal, William J.   II. Pilkey, Orrin H.,
1934–   . III. Title.   IV. Series.
TC225.L59M38   1984   333.91′716′0974721   83–20670
ISBN 0–8223–0501–1
ISBN 0–8223–0502–X (pbk.)

To all those who wish to preserve the meeting place of land and sea for tomorrow's children

Contents

Figures and tables

Figures

Tables

Foreword

The south shore of Long Island is one of New York's greatest recreational assets. On hot summer days hundreds of thousands of people can be found frolicking on Coney Island, Rockaway Beach, or Jones Beach, and the development associated with this is extensive. At the same time there are also stretches that are only lightly developed. On Fire Island National Seashore one can still find secluded beaches and natural dune fields.

The south shore, however, is eroding—much of it at around 6 feet per year. Houses and beach facilities that were built back from the shoreline years ago are now protruding into the water. This erosion is not a recent phenomenon, and over the years the question of what to do about it has been asked many times. Frequently the answer has been to put up some sort of artificial structure like a seawall, groin, or revetment or to replace the sand that has disappeared with sand taken from somewhere else. We now know that such measures, known ironically as shoreline stabilization, not only fail to halt erosion but actually increase it. And it is not just the developed areas that it was hoped such measures would protect that are affected. The attempts to stabilize the shoreline of developed areas have caused erosion to increase in lightly developed ones.

So the question of what to do remains, and it has become more urgent because development of the south shore is continuing.

This situation is not unique to New York. All of America's coastal states, particularly those on the East and Gulf coasts, face these problems. Some years ago we took part in writing a book that dealt with this situation in North Carolina. Entitled *From Currituck to Calabash: Living with North Carolina's Barrier Islands*, it was published originally by the state government. *Currituck to Calabash* has had, we feel, a significant impact on coastal development in the state, and its success led to support from two federal agencies to produce a series of similar books for other states. Eventually there will be a volume for every coastal state.

The goals of the series, which is being published by Duke University Press, are those of the original book: to provide people with information on shoreline erosion and other coastal processes, as well as with site-specific analyses, so that they can make reasonable and informed decisions, both public and private, about what to do now at the shore.

As a supplement to the series we have written another book that expands on an important subject covered in each of the state books—the best way to build at the beach. It is entitled *Coastal Design: A Guide for Builders, Planners, and Home Owners*, and it has recently been published by Van Nostrand Reinhold.

The present book is the fourth in this series. In line with our policy of having the principal author of each book be a prominent coastal geologist from the state, we have recruited Dr. Larry

McCormick to write the book on the south shore of Long Island. When we asked around the geologic community for suggestions about the most qualified person to do this book, Dr. McCormick's name was at the top of everyone's list. He is eminently qualified for the task of compiling the available data, culling them, and presenting them in a form accessible to the nonscientific public. We feel very fortunate in having "captured" him. Dr. McCormick received a Ph.D. degree in coastal geology at the University of Massachusetts. Since that time he has taught at Southampton College and has been involved in the political and scientific aspects of the New York shoreline decision-making process. He is a familiar figure to those who attend public hearings on the future of New York's coast.

His work was sponsored by the New York Sea Grant Institute under a grant from the Office of Sea Grant, National Oceanic and Atmospheric Administration, U.S. Department of Commerce. Accordingly, the U.S. Government is authorized to produce and distribute reprints for governmental purposes notwithstanding any copyright notation appearing in this volume.

The overall project of producing these books is an outgrowth of initial support from the National Oceanic and Atmospheric Administration through the Office of Ocean and Coastal Resources Management. The project was administered through the North Carolina Sea Grant Program. We have recently received support from the Federal Emergency Management Agency to expand the book project to all coastal states. The technical conclusions presented herein are those of the authors and do not necessarily represent those of the supporting agencies.

We owe a debt of gratitude to many individuals for support, ideas, encouragement, and information. Peter Chenery of the North Carolina Science and Technology Research Center and Richard Foster of the Federal Coastal Zone Management Agency gave us encouragement and support at critical junctures of this project. Doris Schroeder has helped us in many ways as a Jill-of-all-trades over a time span of more than a decade. Dennis Carroll, Jim Collins, Jet Battley, Peter Gibson, Gloria Jiminez, Melita Rodeck, Richard Krimm, Chris Makris, and especially Mike Robinson helped us through the Washington maze. We are in the debt of the many coastal residents, fellow geologists, coastal engineers, and state and local government officials—too numerous to name—who enthusiastically provided us with a wealth of data, ideas, and "war stories."

Special recognition belongs to Betty Glendenning for typing and retyping revisions from four different authors. Clint Myers and Barbara Gruver were responsible for putting together most of the line illustrations. The second field trip guide, to Democrat Point and Jones Beach barrier island, was written by Fred Wolff and Robert Johnson.

Thanks are due Peter Sanko of the New York State Sea Grant Organization for his help in providing the authors with invaluable information. In addition, Gilbert Nersesion of the New York District of the U.S. Army Corps of Engineers was instrumental in providing the authors with district reports that helped greatly. Neal Bullington of the National Park Service at Fire Island National Seashore helped in a variety of ways with the Fire Island portion of the book.

Orrin H. Pilkey, Jr.
William J. Neal
series editors

Living with Long Island's south shore

1. Historic trends and occasional catastrophes

It was a fall morning in 1938 on Long Island when a tropical storm was reported off Cape Hatteras, North Carolina, at 7.00 A.M. The Burghards of Westhampton Beach planned a trip into town from their beach-front home on Dune Road to see a tennis match. At 10:00 there was a strong north wind blowing, which shifted to the northeast around 11:00. Fall is a time when northeasters are expected, so they canceled their tennis plans and decided instead to spend the afternoon at home and watch the surf build. About this time Carl and Selma Dalin, who cared for the Burghard home, and Mr. Burghard noted water seeping into the basement from the floor. Apparently the building surf had saturated the sand, and water was being forced into the basement of their home. The wind had increased, and it looked like this northeaster would be a good one.

About 1:00 P.M. their neighbors' 30-foot boat began to drag its mooring, and George Burghard called the Coast Guard. At 1:30 George secured an antenna guy line and much to his surprise was lifted off his feet by the increasing gale. A friend, who had planned to spend the afternoon at the Burghards' watching the storm, phoned to say he wouldn't be coming because his garage had just blown away. A great deal more than one garage was destined to blow away that afternoon.

Mrs. Burghard had been sewing next to a window and moved her material to a safer position just before the window caved in, showering the room with glass, driving rain, and wet sand. As they struggled to cover the shattered window the storm intensified, and at 2:30 the first waves broke over the top of the dune. Thirty minutes later white water boiled through the downstairs hallway. By the time a lone Coast Guardsman arrived at the Burghards' home at 3:30, the downstairs was filled with white water and every wave was cresting the dune line. It was clear that they would be required to evacuate their home, but the Dalins were physically unable to follow. Mr. Dalin was sitting clasping a fence post as the surf surged about him, and Mrs. Dalin leaned against a nearby pole for support.

At 4:15 the wind shifted to the southeast and a flood of green water came over the dunes in surges. By then the waves were heavily armed with wreckage that could kill in an instant. The threatened residents could never walk to the bridge and reach the safety of the mainland, so they decided to ride debris across the bay to safety. There was no sound except that of the storm, the low pitch of the surf, the shrill whine of the wind, and an organ-like note that filled the intervening scale. Their neighbor's home disintegrated next to them without a sound.

The Burghards made it safely across the bay in the company of the Coast Guardsman in what can only be described as a heroic voyage aboard a large piece of flotsam. The remains of the Dalins were located later among the wreckage washed across the bay.

Similar events were recorded at many other localities along the south shore. On Fire Island 127 houses were destroyed in the community of Saltaire, 91 at Fair Harbor, and 29 at Oak Beach. Along Dune Road at Westhampton Beach only 26 homes were left from a prestorm total of 179 (figs. 1.1 and 1.2). The dunes were swept into the bay, and the barrier island was breached by channels along much of its length. The water submerged many south-shore communities, reaching a depth of 6 feet in downtown Westhampton Beach. So much salt spray filled the air that trees were defoliated as much as 5 miles inland.

At the same time that this disaster was in progress, the city of New York recorded only gale-force winds and some limited flooding. The city had difficulty realizing the scale of the damage wrought by the storm on their neighbors less than 50 miles to the east. Similarly, there are few people living on the south shore today who appreciate the damage a storm of this magnitude can accomplish in a few short hours. The result is that the current south-shore population lacks the experience to evaluate the danger posed by the threat of an intense coastal storm. As Santayana once said, "Those who cannot remember the past are condemned to repeat it."

There is a very human tendency to think of storms like the 1938 hurricane as a one-time happening or to think that modern technology has somehow lessened the possibility of a repeat of the 1938 disaster. On the contrary, the south shore is probably less ready to withstand the onslaught of a major storm than in 1938, and the number of homes built perilously close to the edge of the sea has dramatically increased. The south shore has the potential for a natural disaster the magnitude of which might rival an earthquake in a California city adjacent to the San Andreas fault line. We know that the catastrophe is on its way. We lack only the knowledge of when it will strike.

Despite the certainty of future disasters and the long record of retreating beaches on the south shore of Long Island, the popularity and desirability of shore-front real estate remains high. Horror stories like the one just told will not discourage the development of our beaches.

The thought that any tale, no matter how graphic, will scare people away from beaches that are beautiful and tranquil most of the time is simply wishful thinking. There are too many people willing to try to beat the odds. Las Vegas and Atlantic City have grown rich on the gambling spirit that drives people to test chance. The authors recognize this fact of life, and in this book we attempt to give the reader a better understanding of how the beaches of the south shore behave. Our hope is that we might discourage the most prudent from building on the water's edge and guide the remaining adventurers in a way that will reduce the odds working against them.

No less important is our desire to give casual visitors to the south shore a deeper understanding of the processes at work there in order to increase their enjoyment of it. The result, we hope, will be a more enlightened approach to the management of our coastal resources.

In order to serve these purposes we begin in this chapter with a

Fig. 1.1. Aerial view of Saltaire taken on September 23, two days after the 1938 hurricane. The view is toward Great South Bay from the ocean side of the barrier. Note that significant damage took place a great distance back from the shoreline. In Saltaire alone 127 homes were destroyed. Source: *New York Times*.

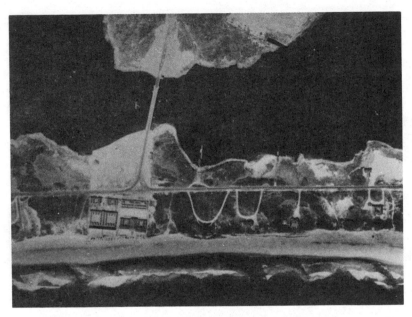 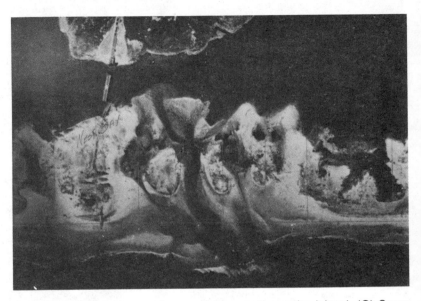

Fig. 1.2. (A) Aerial view of Jessup Lane area in 1928. (B) Same area after 1938 hurricane; note that the bridge to the mainland has been totally destroyed and that water has washed over the entire island. (C) Same area reconstructed and more densely developed in 1972. Source: Suffolk

discussion of the primary reason for long-term beach erosion and then turn to the subject of storms that punctuate the slow retreat of the beaches toward the land.

In the second chapter we explain how the coastal system from Montauk Point to Coney Island operates. The key word here is *system*. Even though the high cliffs and boulder beaches at Montauk Point look and act differently than the sand beaches in Southampton, Long Beach, and Coney Island, they operate as a single system. The narrow strips of sand that separate the ocean from the bays and mainland on the south side of Long Island, and

County Department of Waterways.

which we call *barrier islands*, are simply a westward extension of the system that begins at Montauk Point.

The importance of considering the south shore as an interrelated system is the theme that runs through most of the chapters in this book. Chapter 3 points out what has happened where man has interfered with the system, and in chapter 4 the natural history and erosion rate for each portion of the south shore are considered. Chapter 5 describes where we stand with respect to government management of the system, and chapter 6 is included for all those who decide to disregard the warning not to build on a beach. It describes the best construction techniques for the coastal zone.

Historic trends: rising sea level—the root of the problem

Although cataclysmic storms like the 1938 hurricane are the most spectacular component of the coastal scene, a second and equally insidious hazard confronts the beach-front dweller. It is the slow but persistent rise of sea level, which has been going on for thousands of years but which accelerated about 1930 (fig. 1.3).

The accelerated rise in sea level and retreat of the shoreline is probably related indirectly to the rush-hour traffic on the Long Island Expressway. Each car is burning petroleum and producing carbon dioxide (CO_2). Worldwide burning of fossil fuels is producing vast amounts of this gas. Gradually the level of CO_2 is increasing, causing the atmosphere to retain more of the sun's energy. As the atmosphere warms, the ice caps melt and sea level rises.

The warming of the earth as a consequence of the processes described above is widely known as *the greenhouse effect*, and the National Academy of Science has warned that the acceleration can be expected to continue in coming decades. More south-shore beach homes will be awash in the surf.

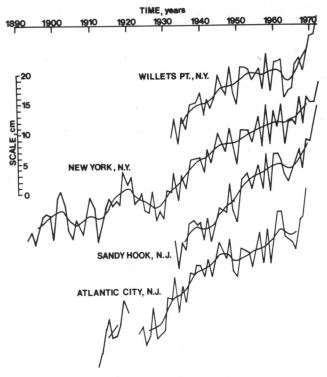

Fig. 1.3. Tide-gauge records from 1890 to 1972 indicate an increase in the rate of sea-level rise about 1930. Source: Steacy D. Hicks, "On the Classification and Trends of Long Period Sea Level Series," *Shore & Beach*, 1972, pages 20–23.

It is important to realize that although the rise in sea level is only about 1 foot per century, the amount of shoreline lost as a consequence is substantially more.

Homes and property that stand in the way of this gradual invasion of the sea must be moved, defended, or lost. It would be helpful if the advance of sea over the land were perfectly regular so that we could predict exactly when the fence in the front yard will be lost, and how long it will be before the front steps will fall. Unhappily, the retreat is erratic. You lose 30 feet of sand dune in front of your house while the fellow down the beach loses none. For 5 years your dune is untouched, then in a series of spring storms you lose another 30 feet.

The Wainscott home shown in figure 1.4, about to fall onto the beach, once had a hedge and a front yard between the beach and the front door. In only 5 years the yard melted away. It narrowly escaped destruction in an alleged insurance fire, only to be leveled later by bulldozers as a hazard to public safety. The beach in front of this property has since widened, the dune has been partially reconstructed, and only 3 years ago a prospective buyer of the property asked the authors if it was a safe construction site.

Eastern Long Island beaches have been retreating at a rate of about 1 foot per year except where the works of man have substantially altered this rate. This figure has a limited utility, however. It is an average reached over a long period of record for a long expanse of beach. Like any average, it is useful for looking at long-term trends, but it doesn't mean much for the individual property owner. The person that assumes that the 20 feet of dune ridge in front of his home will protect him for 20 years is apt to get an unpleasant surprise.

Storms

Storms are the most spectacular of the events that occur at the shore—certainly more spectacular than the very gradual sea-level rise. The most obvious aspects of storms that most people think about are the wind and the waves. High winds damage homes that are poorly built or which have not been "shuttered up" or otherwise prepared for the storm. It has been repeatedly observed in storm postmortems that shutters, roofs, doors, and even lawn furniture may blow into adjacent buildings and act as battering rams. Waves chew away on the beach and cause rapid shoreline retreat, undermining houses—including those that are well built. A third aspect of storms and one that should never be ignored is the *storm surge*. This is the rise in the level of the sea during the storm. It can be responsible for awesome damage. Because of the storm surge, waves no longer break on the beach, they break on the inland dunes or the nearest buildings. In Mississippi in 1969, Hurricane Camille produced an unusual storm surge in which the level of the sea at the shoreline was 20 to 30 feet higher than normal. A 30-foot storm surge should give anyone who owns property near the 10-foot elevation pause for reflection.

Storm surge is produced by the shoreward-pushing action of the wind and by the low atmospheric pressure of storm systems. The low atmospheric pressure causes water to bulge upward,

Fig. 1.4. Home in Wainscott immediately east of the end of Beach Lane about to be claimed by the sea.

simply because there is less weight from the atmosphere pushing down on the water surface.

It is important to understand some basic facts about storm surge. First, its effect can be magnified or diminished by the tides.

On Long Island beaches, a tidal movement sweeps by twice each day, causing two high tides and two low tides in about 25 hours' time. These are the changes in water level produced by the proximity and position of the sun, moon, and earth. When the three are lined up, they produce a maximum effect resulting in the highest high and the lowest low water each month. These are called *spring tides*, and they happen twice a month, coinciding approximately with the new moon and the full moon.

Obviously a storm-related surge in water level that occurs co-incident with a high spring tide, especially if it happens while the water is rising, carries with it a greater potential for flooding than if the surge were coincident with a spring low. The 1938 hurricane described earlier, one of the most destructive storms in history, arrived at the south shore on a rising spring tide (fig. 1.5). According to calculations made by the U.S. Army Corps of Engineers, the surge for this storm was about 9 feet.

Another important fact to understand about storm surge is that different kinds of storms create different kinds of surge. There are two kinds of storms that strike the south shore. The first are the northeasters that develop along frontal zones separating major air masses. The second are the major cyclonic storms that are spawned in more southerly latitudes and which roar northward as hurricanes.

In the case of the extratropical or nonhurricane storms (usually northeasters), the surge occurs over the period of a day or two, and the high surge is followed by a period of negative surge (lower than expected tides). This type of surge is well represented by the records of a storm on February 19 and 20, 1972. The upper portion of figure 1.6 shows the actual changes in water level that occurred, influenced by both the astronomic tides and the storm surge. The lower portion of the figure has the predicted astronomic tides of the 1972 storm subtracted out so that you see only the influence of the storm surge. Thus, this surge reached a maximum of 4 feet above normal sea level and lasted a day and a half. The waves that ride atop these combined tides and storm surges gnaw away at the beaches and dunes, producing the erratic retreat of the shore mentioned earlier.

The surge associated with hurricanes is markedly different from that of a northeaster. The rise in water level is much more rapid, short-lived, and often more spectacular (fig. 1.7), and it is followed by a series of smaller surges. The first surge for a major hurricane usually lasts only 2 or 3 hours.

Tide gauges are specifically designed to measure slow-moving changes in water level such as the tides or storm surges, not to measure the rapidly moving waves of the surf that break on our beaches. In a hurricane these waves of the surf zone may be 10, 20, and even 30 feet high. Half that wave height must be added to the still-water level of the storm surge because half the wave height is above still level and half below. Add to this an unknown additional height for the distance waves run up the beach after they

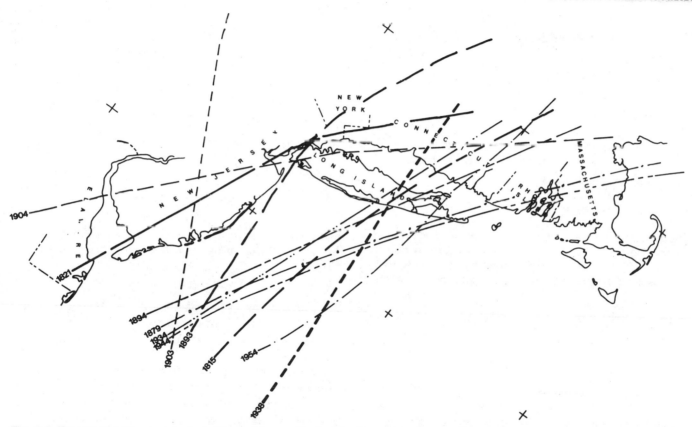

Fig. 1.5. Tracks of major tropical storms that have affected the Long Island area. The sheltering effect of the New York Bight causes most storm tracks to cross the eastern end of Long Island.

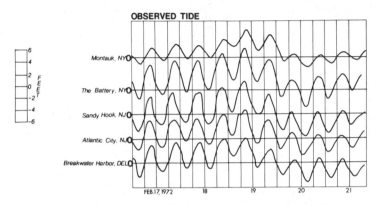

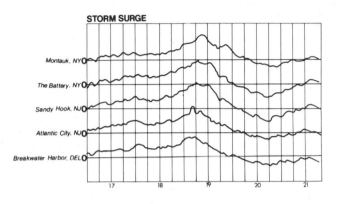

Fig. 1.6. The graph at left shows the daily rise and fall of tides superimposed on the rise in water level caused by a storm surge occurring on February 17, 1972. The graph at right shows the storm surge with the daily tidal oscillation removed. Source: figure 8 in N. Arthur Pore and Celso S. Barrientos, *MESA New York Bight Atlas Monograph No. 6: Storm Surge* (Albany, NY: New York Sea Grant Institute, 1976).

break and you have the total height to which water damage can be expected:

tide height + storm surge + ½ wave height + wave run-up =

total height affected by waves.

It is not difficult to see why even a high dune ridge is no bastion of safety in a severe hurricane.

The 1938 hurricane illustrated an interesting aspect of hurri-
canes on the south shore of Long Island. It showed dramatically that the area east of the eye of the storm will suffer the greatest damage. The people in New York City did not even know a hurricane had struck Long Island.

If a hurricane is to strike from the open sea, it must approach from the south, owing to the east-west orientation of Long Island and the presence of the New Jersey shore to the west. The winds in the storm always rotate counterclockwise so that the winds on the east side of the eye approach from the open ocean where they can attain their highest velocity (fig. 1.8). In addition, winds on

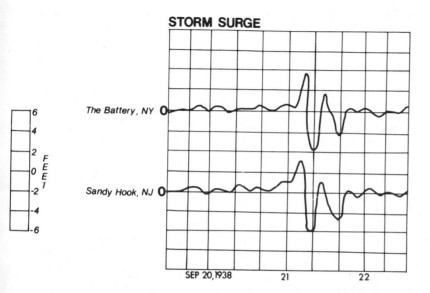

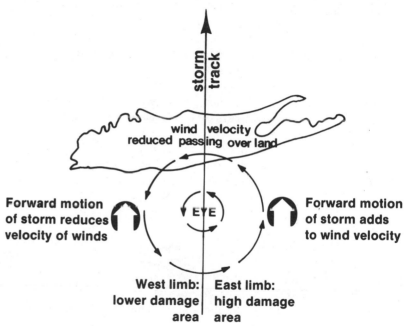

STORM SURGE

Fig. 1.7. Storm surge associated with 1938 hurricane. The length of the surge is much shorter for hurricanes than for longer-lasting northeasters. Source: figure 25 in N. Arthur Pore and Celso Barrientos, *MESA New York Bight Atlas Monograph No. 6: Storm Surge* (Albany, N.Y., New York Sea Grant Institute, 1976).

Fig. 1.8. The geometry of a hurricane causes the east limb of the storm to have a higher velocity than the west limb.

the east limb of the storm are intensified by the forward-moving velocity of the storm, whereas those on the west limb tend to be canceled out by the forward movement of the storm. For example, if winds in the hurricane have a rotational velocity of 100 miles per hour and the storm is moving north at 60 miles per hour, then east of the eye the winds will be 160 miles per hour while those on the west will be only 40 miles per hour. It is not surprising that New Yorkers failed for hours to realize the disaster that had struck eastern Long Island.

We strongly advise that you do not wait around to see which half of the hurricane will strike you. If it appears that a hurricane is headed your way, take action (see appendix A).

Assessing the risk

Any time you are going to the beach, whether permanently as a resident or even just temporarily as a visitor, you should be aware of the risk of a dangerous storm. You should assess the risk carefully. One measure of the degree of risk involved is the frequency of storms (table 1.1): the higher the frequency of storms, the higher the frequency of trouble.

Another factor to consider in assessing the safety of a coastal location is the amount of time you have to prepare for a storm or to evacuate in advance of one. Most of us are accustomed to the slow progress of tropical storms that we hear about each year approaching Florida or the Gulf of Mexico. This creates the false impression that a hurricane hundreds of miles distant from our

Table 1.1. Estimated storm frequency

	Period of record	Number of occurrences	Frequency per 100 years
Unusually severe	1635–1956	9	2.8
Severe	1801–1956	14	9.0
Moderate	1901–1956	13	23.2
Threatened Long Island	1901–1956	36	64.3

shores will allow plenty of time to batten down the hatches and get the family to a place of safety. In fact, it is common practice for hurricanes to accelerate their forward movement as they progress northward. Remember, the damaging 1938 storm was reported off Cape Hatteras at 7:00 A.M. on the morning of September 21. It crossed Long Island only 8 hours later. The forward progress of the storm was in excess of 60 miles per hour! Always assume you will have less escape time than indicated by weather bureau predictions of storm landfall.

Select an escape route ahead of time. Check to see if any part of it is at a low elevation, subject to blockage by water from overwash or flooding; if it is, seek an alternate route. If no alternate route exists, allow more time for flight. Note whether there are bridges along the route. Remember that some residents will be evacuating pleasure boats, and that fishing boats will be seeking safer waters; thus, drawbridges will be accommodating both boats and automobiles.

Reevaluate the escape route you have chosen periodically—especially if the population of the area in which you live has grown. With more people using the route, it may not be as satisfactory as you once thought it was.

The Suffolk County Department of Emergency Preparedness (see appendix B, under Civil disaster and preparedness assistance) has been designated to coordinate aid and assistance in the event of a natural disaster, but the efforts of this office will have their greatest effect after the storm. The first decision, to stay or evacuate, is up to the individual. That decision is best made by keeping tuned to a local radio station.

The National Oceanic and Atmospheric Administration has made a study of barrier-island evacuation problems. They estimate that many of the most heavily developed islands cannot be entirely evacuated. Presumably this is because many people will not move until the winds are actually blowing.

Use the escape route early. Be aware that some islands have limited routes for escape to the mainland. In the event of a hurricane warning, leave the island immediately; do not wait until the route is blocked or flooded. Anyone who has experienced the evacuation of a community knows of the chaos at such bottlenecks. Depend on it: excited drivers will cause wrecks, run out of gas, have flat tires, and cars of frightened occupants will be lined up for miles behind them. Be sure to have plans made about where you will go. Keep alternative destinations in mind in case you find the original refuge filled or in danger. Otherwise you should start muttering the salute given to Caesar by the Roman gladiators as they entered the arena: *Morituri te salutamus*, which means, "We who are about to die salute you."

Hurricane Carmen, which hit the Gulf Coast in September 1974, illustrated the desirability of leaving early to miss the traffic jam. Over 75,000 people are said to have evacuated from what were thought to be the danger areas in Louisiana and Mississippi. The traffic was bumper to bumper on the few roads leading north. One accident backed up traffic for 19 miles. Motel lobbies were filled with people looking for a place to stay; all rooms were taken. Weary people were forced to continue traveling north until they found available space. Similar experiences should be anticipated for New York, because island populations have grown enormously since the last major hurricane.

In closing this discussion we might note another phenomenon that occurs in every storm. There are the adventurous souls who rush to the island to experience the excitement of the storm. One island dweller bragged to us that he and his 8-year-old son were the last to get *on* Sea Island, Georgia, just before Hurricane David brushed by in 1979. A Fire Island resident talked to us with bravado of the superb vodka gimlet party he hosted as bay water gradually crept over the floor of his house. It seems that rational behavior often gets carried away with the screaming winds.

2. Shoreline dynamics

Long Island presents two distinct faces to the never-ending energy of the Atlantic Ocean. From Montauk Point to Southampton, the south shore of the island is directly exposed along a stretch of coast called *headland* (fig. 2.1). From Southampton west, however, there is a series of narrow barrier islands paralleling and protecting the south shore (fig. 2.2). The beaches that occur on these two settings are among the most dynamic strips of land anywhere.

They are also among the most valued. Every year thousands of us make use of them as visitors or residents. If we are not careful, however, we can do serious damage to the south shore. If we are not careful, we can put ourselves and our possessions in serious danger. To avoid this we must realize that the ocean, headland, barrier islands, and beaches are major elements in a natural system. It is important that we know and understand all the basic elements of this system and how they interrelate.

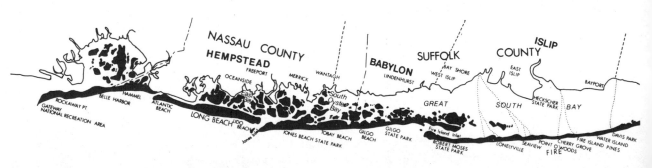

Fig. 2.1. The south shore of Long Island, New York.

Beaches

The beach is an active strip of sand that extends from the base, or toe, of the dune out to a water depth of about 40 feet. It is best understood by looking at it from two separate points of view. One is the map or plan view. This is the one you get when observing the beach from an airplane. The second is cross-sectional. In order to visualize the cross section you must imagine the silhouette created by a plane that slices through the beach perpendicular to the shoreline and at a right angle to the ground. The resulting cross section is called the *beach profile*.

The beach in profile

Beaches experience rapid seasonal changes in the shape of their profile as a result of the changes in wave energy expended on them. In the summer, gentle swells sweep sand upward onto the beach, and it builds up to form a *summer profile* (fig. 2.3). During storms, however, high, steep waves crash onto the beach, and sand is carried back into the sea. The beach flattens. This flattening during a storm is fortuitous because it serves to dissipate the energy of the breaking storm waves over a larger surface area, thus diminishing their erosional effect. Once the storm has passed, the fair-weather waves that follow will slowly return the sand that

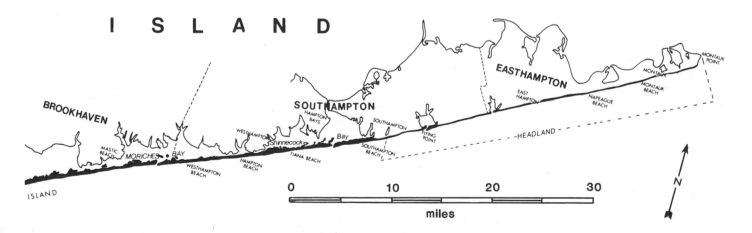

Fig. 2.2. View north across the barrier island at Westhampton Beach. This is the same area shown devastated by the 1938 hurricane in figure 1.2.

has been removed. This rebuilding is a process that may take months or even years.

Even though the conditions for any beach change seasonally to produce alternating summer and storm profiles, there is an average shape for the year. This is the *equilibrium profile* (fig. 2.3), and it should be about the same from year to year. Furthermore, if con-

ditions from year to year are similar, it should remain in about the same position.

The equilibrium profiles of the beaches of the south shore of Long Island, however, are shifting landward. This shift, which has been going on for about 16,000 years, is in large part the result of the rise in sea level discussed in chapter 1.

Many factors affect the rate at which this shift is occurring. One, of course, is the rate of sea-level rise, and another is sand supply. This can be best discussed from the plan point of view.

The beach in plan view

Because beaches are constantly losing sand, how much they are simultaneously adding is very important. If more is lost than added, then a beach may, depending on other factors, shrink, shift landward, or disappear altogether.

Looking at the south shore from above, we can point to two primary sources of sand for its beaches. The first is the headland from Montauk Point to Southampton. Here sand is eroded from glacial cliffs (fig. 2.4). Sand is also brought in all along the south shore from the shallow ocean floor seaward of the beaches.

Sand from both sources is caught in a *longshore current*. This current moves parallel to the beach in the turbulent zone between the point that waves begin to break and the highest reach of the swash of the wave on the beach. The current occurs because most waves approach the beach at a slight angle; this causes a portion of the energy of the breaking waves to be directed along the beach. Sand and other material carried in the longshore current are

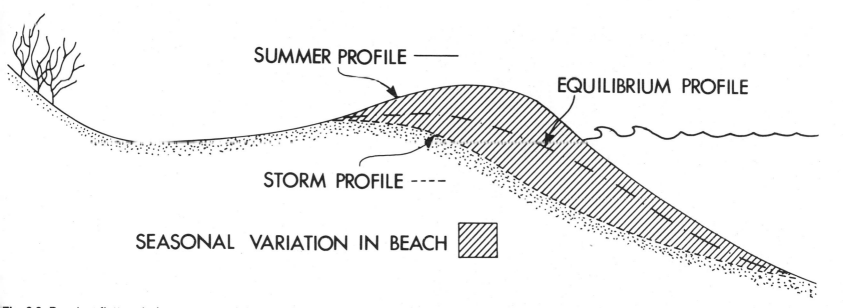

Fig. 2.3. Beaches flatten during storms and then steepen and build outward during quiet periods. Between these two extremes lies the average position of the beach.

referred to as *longshore drift* (or sometimes *littoral drift*). The direction and intensity of its movement vary as wave height or direction change, but in general more material is moved west along the south shore than east (fig. 2.5). In fact, the westward drift actually grows stronger the further west along the shore you go. This is because waves coming from the east are generally stronger than those coming from the west. Common waves are caused by wind blowing across the surface of the ocean. Three variables that affect wave strength are the velocity of the wind, the amount of time the wind pushes a wave, and the area, or

Fig. 2.4. Eroding bluff on the headland at Montauk Point. The bluff has been terraced in front of the home to slow the inexorable advance of the sea.

fetch, over which the wind has blown. Waves approaching the south shore from the west are therefore usually weaker than those approaching from the east because the fetch is less.

Figure 2.6 shows diagrammatically how sediment moves along the south shore. At the eastern end of Long Island, sediment is swept in from offshore, but the longshore drift moves the material off to the west more rapidly than it is being supplied. These beaches are experiencing a net loss in the total amount of sand available to them, and they press against the headland. Westward from Southampton, the sediment contributed by longshore drift and offshore sources is equal to the amount of material being removed.

The westward movement of sand from Montauk to New York City stops in the New York Bight. Here the sand is dumped and comes to a "permanent" resting place. Along the way, however, some sand from the longshore-drift system may be lost through inlets such as Fire Island Inlet.

Interrupting the movement of sand can have devastating effects on the beaches of the south shore. See chapter 3 for a discussion of how we have interfered with this system in the past and what the results have been.

Barrier islands

Thus far we have seen how dynamic beaches are. They react to seasonal conditions by changing the shape of their profile; they depend upon a continuous supply of sediment to replace the sedi-

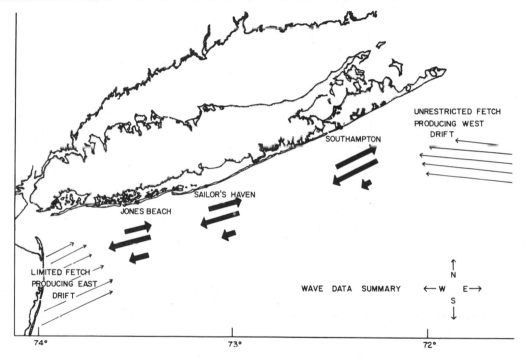

Fig. 2.5. The dynamics of longshore drift for the south shore of Long Island. The top arrows at Jones Beach, Sailor's Haven, and Southampton show the relative effect of waves from the west in moving sediment, and the middle arrows show the relative effect of waves from the east in moving sediment. The lower arrows indicate the increasing dominance of westward longshore drift. Source: Larry McCormick and Margurite Toscano, "Origin of the Barrier System of Long Island, N.Y.," *Northeastern Geology*, 1981, volume 3, pages 230-34.

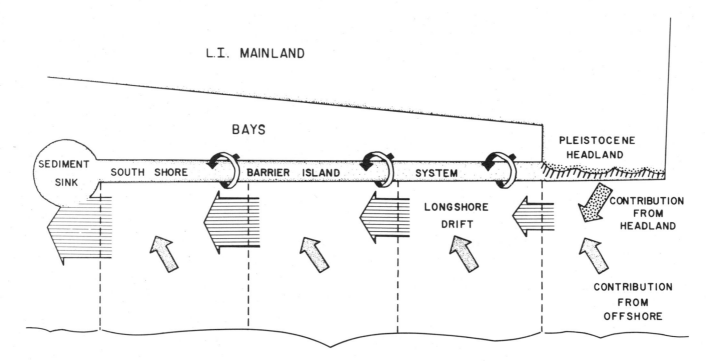

Fig. 2.6. Model for sediment transport along the south shore of Long Island. Source: Larry McCormick and Margurite Toscano, "Origin of the Barrier System of Long Island, N.Y.," *Northeastern Geology*, 1981, volume 3, pages 230-34.

ment they are continuously losing; and they are responding to the worldwide rise in sea level by moving landward. It follows that barrier islands, because a large part of any barrier island is its beach, are also dynamic. But this is not the entire story. Barrier islands are dynamic as a whole.

They have to be. For if an island were static while the beach on the side that faces the ocean—the front side—were moving landward, then the island would eventually disappear. The side of the island facing the bay that separates it from the mainland must also move landward. It grows landward. These two events—the erosion of the front side and the growth of the back side—are the key aspects of what scientists call *barrier island migration*: the movement of barrier islands toward land.

The growth of the back of an island is accomplished in many ways. For instance, wind carries sand from the beach into the island where it is trapped by vegetation; dunes are gradually built up. Also, during storms onshore winds may cause water to wash over dunes at low spots in the dune ridge. The overwash carries sand through the dunes and builds out low, flat aprons of sand into the bay. These are *overwash fans* (fig. 2.7). Eventually these can build up enough to be stabilized by salt-marsh grasses. The island has thus widened.

The process of migration where material is continually moved from the front to the back of the island has been called *roll-over*, and there is evidence that it is occurring on south-shore barrier islands (fig. 2.8). At Cooper's beach in Southampton, extensive beds of peat can be observed cropping out on the beach after

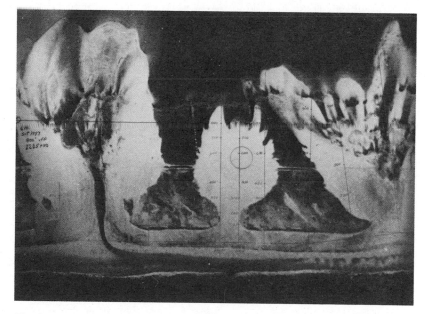

Fig. 2.7. Three overwash fans created on Long Island during the 1938 hurricane. Source: Suffolk County Department of Waterways.

severe storms have removed sand. There is no doubt that this peat was formed in marshes behind the barrier island and that now the barrier has rolled over it, exposing this ancient deposit. The back is now at the front. Similar deposits have been found in borings on Fire Island, and one peat deposit exposed on the beach at

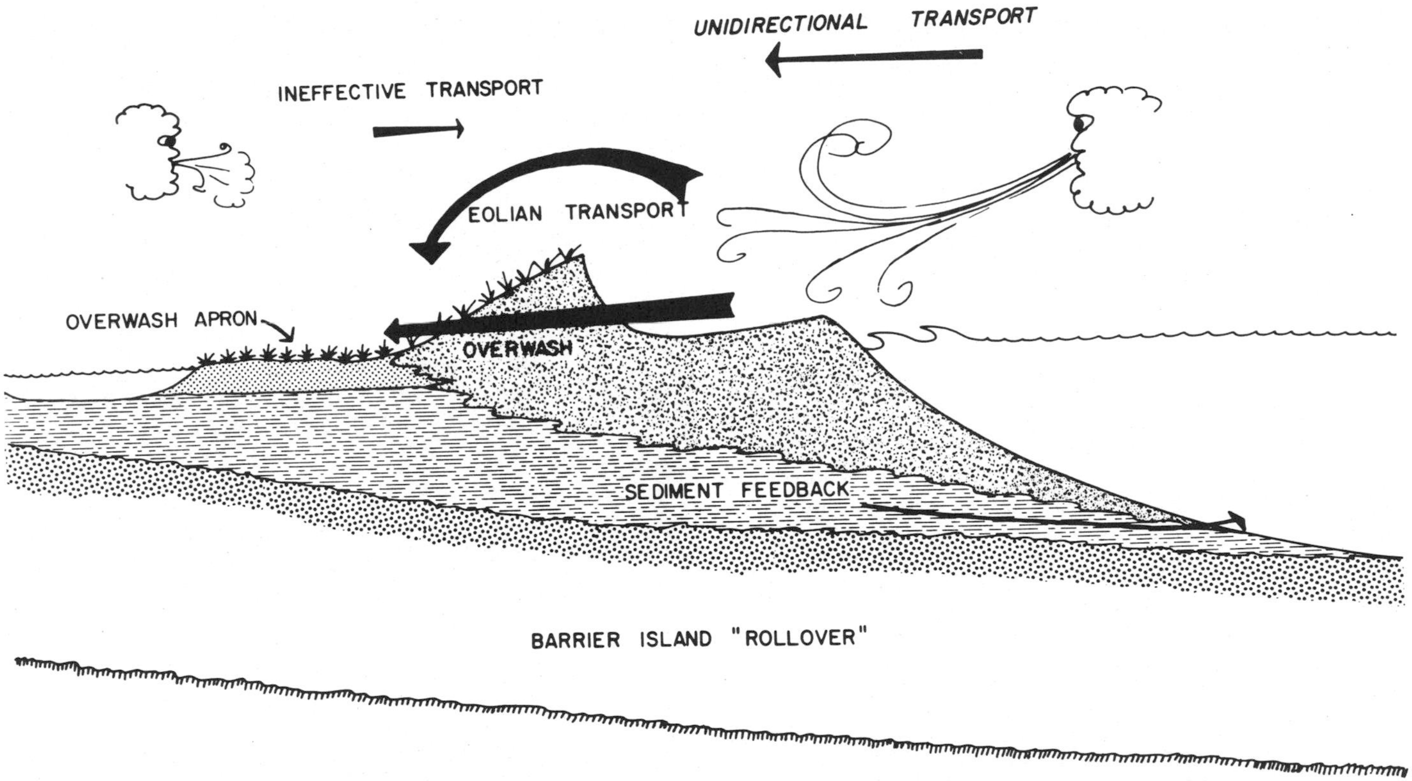

Fig. 2.8. Barrier island roll-over. Onshore winds and waves move sediment from the front to the back of the island so that it rolls over itself. Offshore winds do little to move sediment the other direction. Source: Larry McCormick and Margurite Toscano "Origin of the Barrier System of Long Island, N.Y.," *Northeastern Geology*, 1981, volume 3, pages 230-34.

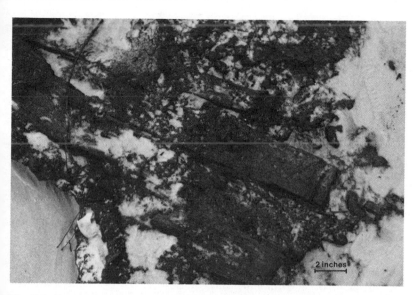

Fig. 2.9. Barrel staves preserved in a thin peat layer exposed on the ocean beach near Democrat Point at the west end of Fire Island.

Democrat Point had formed so recently that barrel staves, once discarded on the back of the island, were imbedded in it (fig. 2.9).

If roll-over is interrupted, for instance by the construction of seawalls (see chapter 3), the results can be serious.

Barrier islands migrate for the same reason that beaches shift landward: the sea level is rising. In fact, rising sea level not only affects barrier islands today, it also played a part in their creation.

The origin of the south shore: a theory

There are several theories on the origin of barrier islands, and there has been much scientific debate. The most widely accepted theory is the one advanced by the late John Hoyt of the University of Georgia and later modified by Donald Swift of the National Oceanic and Atmospheric Administration. Hoyt and Swift suggest that about 16,000 years ago when sea level was much lower than today, and the shoreline was far out on the continental shelf, the action of waves formed beaches that provided sediment for the onshore winds to heap into dunes. As sea level began to rise, the area behind the beach dune ridges filled with water, creating the bays that separate barrier islands from the mainland. As the sea continued to rise, the shore of the mainland retreated and the islands moved landward. With a little additional supposition, the theory of Hoyt and Swift can be applied to Long Island.

About 16,000 years ago the leading edge of the last great continental ice sheet had deposited the mound of rubble that is now the ridge running from New York City to Montauk Point. South of the mound there was a flat accumulation of sand and gravel that sloped gently out to the lowered sea. There is always some melting associated with glaciers, and this *outwash plain* had been built from the sediment deposited by streams draining the ice sheet. The deposits of individual streams were fan-shaped, but eventually these grew together. The shapes of individual *outwash fans* are still plainly visible today along the landward margins of south-shore bays.

A beach dune ridge had been built at the edge of the sea.

The irregularities that once existed along the shore had been smoothed by wave action—the tips of the outwash plain had been trimmed off, and short peninsulas of sand, called *spits*, were formed across the mouths of the streams. The shoreline must have resembled the beaches that exist today between East Hampton and Southampton.

At this point, the melting of the glaciers increased markedly, and the sea, fed by water released from them, started to rise rapidly. It is an open question whether, in the face of this rapid drowning, the dunes, beaches, and spits were immediately separated from the main part of Long Island as true barrier islands or if they were kept pressed back against the outwash plain. In either case, the shoreline retreated landward a great deal between 16,000 and 8,000 years ago. Some authorities feel that the south shore evolved into a true barrier system only when the rate of sea-level rise slowed about 8,000 years ago. One line of reasoning supporting this view is that the balance between sediment lost and sediment gained, which is required for maintenance of the barrier islands, might have been more easily attained during a slower rise in sea level.

Whether early or late, when the balance was attained, the beach profile began moving landward at a rate slower than the outwash plain was being drowned. This was the birth of the south-shore barrier island system that has gradually migrated landward over the following centuries. Only at the eastern end of Long Island has the beach dune ridge been kept pressed against the main portion of the island.

The processes that consume eastern Long Island to maintain the barrier island beaches of western Long Island have been in operation for thousands of years and can be expected to continue into the future. Montauk Point will be destroyed. East Hampton will become the new land's end so that all features we now recognize will shift westward as well as landward. These major changes will not occur within our lifetimes, but the perspective that this picture supplies is useful when considering construction or management problems in the coastal environment. The short-term changes that we will see and that will affect our planning are simply small steps in the overall march of the sea over the land.

The south shore of Long Island: an integrated system

The description of the south shore on the preceding pages should make it clear that the whole shoreline is connected through the mechanism of sediment transportation. A change in any part of the system results in an adjustment somewhere else. Dam up the littoral drift, and downdrift beaches starve. Stop erosion on the cliffs at Montauk, and greater erosion will occur somewhere else. Stop overwash and dune processes from operating, and barrier island migration cannot be maintained. Mine sand for beach replenishment from offshore areas, and erosion will follow. In a system so precariously balanced and so sensitive to change, we must act with caution. The next chapter discusses in more detail some of our past actions and what the results have been in the south-shore barrier island system.

3. Man and the shoreline

Over the years people have responded in many ways to the receding shoreline. You should be aware of what methods have been used and how successful they have been. In general, responses have fallen into two categories: structural and nonstructural.

Structural responses

Structural responses all involve doing something physically to the shore. Because their purpose is to hold the shore in place and keep it from moving, they are often referred to as *shoreline stabilization*. Because they must be designed and built for the most part by engineers, they are also referred to as *shoreline engineering*. The following structural responses are discussed in order of their environmental effects—beginning with the most damaging.

Groins and jetties

Groins and jetties are walls built out into the water perpendicular to the shoreline. *Jetties* are built at inlets, frequently in pairs, and are intended to keep sand from filling in the channel. *Groins* are similar, but they are not built at inlets. Their only purpose is to trap sand carried in the longshore current. The beach on the side of the groin that faces into the current (the updrift side) is maintained by trapped sand. Unfortunately, a corollary of this is that the beach on the other side (the downdrift side) is starved of sand (fig. 3.1).

Some confusion exists about the effects of groins because engineers once argued, and some still do, that when the area updrift of the groin filled completely with sand, the littoral flow would continue as if the structure didn't exist. This view overlooks the fact that sand passing the seaward end of the groin goes into deep water; it is not available to nourish downdrift beaches until it is pushed back up to the shore by waves. The result is a shadow zone of erosion downdrift of a groin, whether filled or not.

There are two important generalizations that can be made about groins. First, the larger one is and the more effective it is in stabilizing updrift beaches, the greater the erosion on downdrift beaches. Second, groins built at the extreme downdrift end of a littoral system tend to have fewer adverse effects downdrift. Groins have been used extensively along the western barrier island of the south shore (fig. 3.2), but only a few have been built on the eastern end of the south shore. The latter have had disastrous effects on downdrift beaches.

Shore-hardening structures

Shore-hardening structures are all built parallel to the shore. *Seawalls* are placed slightly back from the shoreline so that they

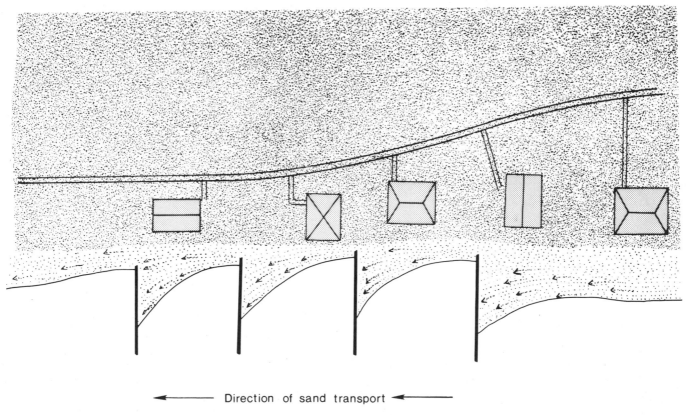

Direction of sand transport

Fig. 3.1. Groined shoreline.

Fig. 3.2. View west along Coney Island showing extensive groin field nearly empty of sand.

receive the full impact of the sea at least once during a tidal cycle. *Bulkheads* are also walls but are set further back—in front of the first dune, or what was the first dune—and are intended to receive only the energy of storm waves. *Revetments* are usually stone facings placed on eroding scarps or bluffs to slow storm-wave erosion. They are frequently made of loose stacks of large stones.

A wave breaking on stone revetments will lose part of its volume in the space between rocks, which reduces the erosive effect of the backwash of the wave.

Seawalls, bulkheads, and revetments all serve basically the same function. They buttress the land to keep it from eroding. The minimum cost of a seawall capable of withstanding, at least temporarily, the waves from the open ocean is of the order of $300 to $600 per linear foot.

The degree to which shore-hardening structures alter coastal processes is open to question. Some believe they cause increased erosion by reflecting more wave energy back on the shore. This also steepens the offshore profile of the beach, which in turn increases the wave energy striking the shoreline. Others say that a properly built structure will avoid these shortcomings. On one point, however, there is no question. They create an immovable wall, and the natural retreat of the beach into the wall causes the beach to narrow or disappear completely (fig. 3.3). Eventually the structure will be undermined and destroyed; in the interim the beach will suffer.

Like groins, shore-hardening structures have been used more extensively on western Long Island. In response to a severe erosion problem, however, individuals in the village of East Hampton have recently constructed revetments along a considerable portion of the village beaches. Beaches subjacent to the walls are already narrow to nonexistent. At the same time prevention of erosion in East Hampton is reducing the sand supply to other south-shore beaches.

Fig. 3.3. Cape May, New Jersey, seawall (1976).

The record of success for groins and shore-hardening structures has been dismal (fig. 3.4). Perhaps the most outstanding example of the abject failure of these structures is found in New Jersey. One of the oldest resort areas in the United States, it felt the need to defend its shores earlier than most areas. Groin followed groin in an attempt to stabilize the shore. Seawalls were built as groins failed, and then the seawalls were raised higher as beaches continued to narrow. Today only the seawalls, groins, and homes that they defend remain. Many of the beaches that attracted the populace and were once so beautiful, including Sea Bright, Monmouth, Long Branch, and Cape May, exist today only in yellowing photographs or the memories of early residents.

Unfortunately it is not necessary to drive to Cape May to observe this "New Jerseyization" in progress. A 1960 report by the U.S. Army Corps of Engineers recommended dune construction, beach fill, and construction of fifty groins along the shore from Fire Island Inlet to Montauk Point. When the storm of 1962 badly eroded the shore at Westhampton Beach, pressure was applied by shore-front interests to have some of the groins authorized in the 1960 plan built in order to stop further erosion. A field of eleven groins was subsequently constructed but was not filled with sand as the Corps had recommended. Following the local government's failure to nourish the groins with sand as it promised, the area immediately downdrift of the field began to erode. In 1967 the ocean overwashed into the bay, and pressure was exerted for the construction of four additional groins (figs. 3.5 and 3.6). The Corps gave in to this request even though their original agreement with the local government had not been honored. The result was construction of the additional groins and transfer of the erosion problem a little further downdrift. Currently there are cries for yet more groins or rebuilding of the beach. Westhampton Beach has taken the first major step toward New Jerseyization.

It is worth noting that agreements were signed at the outset of the project that freed the United States government of claims for damages; this releases the Corps from all financial responsibility for their decisions. The state and local governments are left with any claims for liability. Another interesting footnote to this situation is that no public access is provided to the beaches within the groin field, even though this was a specific condition for its construction.

The Corps claims the project has generated additional problems because state and county governments reneged on promises for beach fill. The state and county are left with liability for damage the project has caused, and the basic problem that the engineers set out to solve persists at a new location. The real problem would seem to be the adoption of a structural solution in the first place.

And that isn't all there is to the story. It isn't just Westhampton that has erosion problems. If viewed from afar, the Westhampton groins can be seen as one of an increasing number of blockages in the south-shore sand supply. Each groin, each seawall, each revetment reduces the sand supply, which results in increased shoreline erosion somewhere else in the system.

Beach replenishment

Beach replenishment involves adding sand to the beach artificially. At first glance such projects would seem to have few negative environmental impacts, but even here there are hidden dangers. Sand to rebuild south-shore beaches has traditionally been taken from the bays or from bars at inlets. With the growing recognition of the value of bay bottoms for the fisheries, engineers have begun to look offshore for sources of sand. A recent survey of sediment in the nearshore zone indicates an abundance of sand with the right size characteristics to be used on south-shore beaches. Mining of this resource has already begun at Rockaway Beach (fig. 3.7), and a similar project is currently planned for the area downdrift of the groins at Westhampton Beach.

Moving sand to beaches from the open sea is a much more expensive operation than mining the bay bottom because the dredges used to mine sand are exposed to waves from the open sea. Furthermore, sand once placed on the shore is not a final solution. It will move off in the littoral drift and must be replaced again and again in order to keep the beach stabilized. Experience in other areas of the country indicates that the erosion rate of a replenished beach is typically at least ten times that of natural beaches. Sufficient money is never available to replenish the entire beach out to a depth of about 40 feet. Thus only the upper beach is covered with new sand: in effect a steep beach is created (fig. 3.8). Off the Connecticut coast, wave patterns changed by a dredged

Fig. 3.4. Effectiveness of a wooden bulkhead in protecting a Southampton home from the incursions of the sea. (A) Bulkhead under attack. (B) Bulkhead destroyed. (C) Home jacked up and moved back from the edge of the water. Photographs A and B by Tom Mills; photograph C by

Larry McCormick.

hole on the shelf quickly caused a replenished beach to disappear. Beach replenishment has been referred to by the U.S. Army Corps of Engineers as an "ongoing" project, but "eternal" is perhaps a better term. Justifying the cost for such a project ($2 million per mile is an approximate figure) is probably possible only in areas where an entire urban environment would be sacrificed if nature were allowed to run her course.

A less obvious danger in mining sand offshore is that we might be cutting off the natural source of sand to the beach. At present many engineers believe that (1) significant amounts of sand do not move about in water depths greater than 30 feet and (2) the net movement of sand is seaward from the beach. If these assumptions were correct, mining sand offshore might pose little threat to the shoreline. The difficulty is that there are data that suggest that both of these assumptions are wrong. The quantities of sand in the littoral drift along south-shore beaches virtually require that some sand be moving onshore, and the shapes of nearshore sand bodies strongly suggest that during storms sand is moving to depths much greater than 30 feet. Add to this the changes in wave patterns that might result from digging a hole in the bottom offshore and we have the ingredients for the kind of failures experienced in the past with more conventional structural solutions. One must keep in mind that beach replenishment is being advocated by the same factions that gave us Cape May and Sea Bright, New Jersey, Westhampton Beach, Long Island, and a multitude of other unsuccessful structural solutions.

Fig. 3.5. View east of groin field at Westhampton Beach.

Fig. 3.6. Erosion of beach and damage to homes in the sand-starved area downdrift (west) of the Westhampton groin field.

The mobile home alternative

We have repeatedly seen homeowners spend small fortunes attempting to protect their homes from the encroachment of the sea with groins or shore-hardening structures—only to see these structures fail and the residents retreat, house and all, as a last resort. In most cases this should be the first option. Although we are accustomed to thinking of houses as immovable objects, it simply isn't so. Moving a structure out of harm's way has no effect on the beach, assures the owner a period of safety, and preserves the beauty that was after all the original attraction.

Structural solutions and decision making at the beach

Shoreline stabilization of any kind, including beach replenishment, is a danger not only for its environmental effects but also

Fig. 3.7. Dredge engaged in beach nourishment operation at Rockaway Beach. Source: U.S. Army Corps of Engineers, New York District.

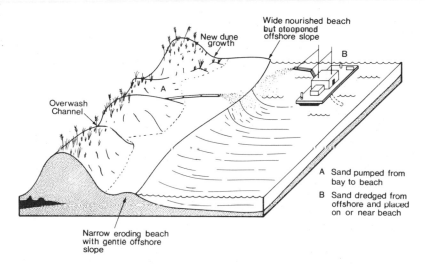

Fig. 3.8. Beach nourishment.

for the role it almost invariably plays in policy making at the beach. Obviously it is carried out initially in areas where enough development has occurred to draw attention to and create concern for erosion. By masking or briefly forestalling the erosion, it encourages development to continue. The result is often the construction of costly real estate projects. The owners of these projects become an increasingly large political faction in favor of even more grandiose stabilization measures to protect their property. Finally, the investment becomes so great that, in effect, alternatives to stabilization no longer exist (for instance, Long Beach and Coney Island). The beach becomes fixed in place despite rising sea level, disappearing sand, and rising public indignation about the use of tax dollars to protect the interests of a relatively small number of people—the shore-front owners. At the same time, public access to beaches is frequently restricted, and the beaches are degraded. It seems that people love houses and they love the beach, but they love their houses more.

Nonstructural responses

Are there ways that we can respond to an eroding shoreline that don't involve trying to stabilize the system? The answer is a resounding *yes*, but implementation requires advanced planning and sometimes politically painful actions.

Response 1: do nothing

Structures that can't be moved—houses, lighthouses, condominiums—should be allowed to fall in when the sea takes them. This is obviously an unpopular response from the standpoint of the beach-property owner, and it places a financial burden on the National Flood Insurance Program (NFIP).

The National Flood Insurance Program covers part or all of the financial loss (see chapter 5) due to flooding. However, it also imposes controls on new construction and places coastal dwellers in a common insurance risk pool. The objective of this approach is to minimize financial loss in the event of disastrous storms. In this sense it performs a function similar to the structural solutions with none of the adverse environmental impacts. Although insurance is not usually considered as a solution for erosion problems, the related construction requirements and actuarial rates may prove to be a cost-effective and ecologically sound way of approaching this nationwide problem when used in conjunction with other programs.

It is interesting to note that south-shore beach property is protected by the federal government in two ways—resulting in a kind of double jeopardy for the taxpayer. Great sums of public money are spent to protect coastal property with groins, revetments, or rebuilt beaches; but if these measures fail, the homeowners are able to collect federally backed insurance for damage to their homes. It is ironic that even when the taxpayer must pay for the damaged house, federal policies allow expenditures of money on stabilization projects that lead to degradation of the taxpayer's beach.

Response 2: set-back lines

One response to erosion is to require that all construction be set back a designated distance from the shoreline. The only problem with this response is that, historically, set-back lines have not been strictly enforced. Furthermore, sooner or later the sea-level rise will catch up with the set-back house. Hence, set-back lines must be as mobile as the shoreline in order to succeed—that is, the set-back line by definition must move back as the shore retreats (and so must the houses).

Response 3: forbid stabilization

Pass a regulation that forbids stabilization of barrier beaches where federal flood insurance is available. This simple alternative recognizes that the taxpayer will pay for the loss anyway, so why should he pay for the high cost of stabilization too? Even private stabilization should be prohibited. A seawall or groin built by private individuals destroys the beach just as effectively as a public structure.

Response 4: the Fire Island solution

Park officials of the Fire Island National Seashore have proposed that the beaches be left in their natural state and that if sand is to be pumped onto the barrier it be pumped to the bay side of the island, thereby aiding the natural island-migration process. This means that on Fire Island many more homes will be destroyed by the sea. Such a solution would also place a greater burden on the National Flood Insurance Program.

Response 5: get involved

Individuals must begin to urge federal and local planning groups to take heed of the warning given by the National Academy of Science regarding the rise in sea level. Research directed toward learning more about how beaches and barriers migrate should be encouraged in the hope that we can learn how best to live with an ocean that is slowly drowning our land. We must insist that planners adopt a policy of living with the inexorable retreat of our shores rather than assuming we live in a world with absolutely fixed boundaries.

Truths of the shoreline

If one looks at the examples of Cape May, Westhampton Beach, and other shoreline areas, certain generalizations emerge quite clearly. These are equally evident to scientists who have studied the shoreline and old-timers who have lived there all of their lives. As aids to safe and aesthetically pleasing coastal development, they should be the fundamental basis of planning.

There is no erosion problem until a structure is built on a shoreline. Beach erosion is a common, expected event, not a natural disaster. Naturally occurring shoreline erosion is not a threat to barrier islands. It is, in fact, an integral part of island migration (see chapter 2). Many developed islands are migrating at surprisingly rapid rates, though only the few investigators who pore over aerial photographs are aware of it. Whether the beach is growing or shrinking does not concern the visiting swimmer, surfer, hiker, or fisherman. It is only when man builds a "permanent" structure in this zone of change that a problem develops.

Construction by man on the shoreline causes shoreline changes. The sandy beach exists in a delicate balance of sand supply, beach shape, wave energy, and sea-level rise. Most construction on or near the shoreline changes this balance and reduces the natural flexibility of the beach. The result is change that often threatens man-made structures. Dune removal, which often precedes construction, reduces the sand supply used by the beach to adjust its profile during storms. Beach cottages, even those on stilts, may obstruct the normal sand exchange between the beach and the shelf during storms. Similarly, structural devices interrupt or modify the natural cycle.

Shoreline engineering protects the interests of a very few, often at a very high cost in federal and state dollars. Shoreline engineering is carried out to save beach property, not the beach itself. Beach stabilization projects are in the interest of the minority of beach-property owners rather than the general public. If the shoreline were allowed to migrate naturally over and past the cottages and hot dog stands, the fisherman and swimmer would not suffer. Yet beach-property owners apply pressure for the spending of tax money—public funds—to protect the beach. Because these property owners do not constitute the general public, their personal interests do not warrant the large expenditures of public money required for shoreline stabilization.

Exceptions to this rule are the beaches near large metropolitan areas. The combination of extensive high-rise development and heavy beach use (100,000 or more people per day) provides ample economic justification for extensive and continuous shoreline stabilization projects. The cost of replenishing Westhampton Beach, for example, is equal to that of replenishing Coney Island, New York, which accommodates tens of thousands more people daily during the summer months. It is more justifiable to spend tax money to replenish the latter beach, since protection of this beach is virtually in the interest of the public that pays for it.

Shoreline engineering destroys the beach it was intended to save. If this sounds incredible to you, drive to Rockaway Beach or New Jersey (or Sea Island, Georgia; South Florida; or Galveston, Texas) and examine their shores. See the miles of "well protected" shoreline—without beaches (fig. 3.9)!

The cost of saving beach property through shoreline engineering is usually greater than the value of the property to be saved. Price estimates are often unrealistically low in the long run for a variety of reasons. Maintenance, repairs, and replacement costs are typically underestimated, because it is erroneously assumed that the big storm, capable of removing an entire beach replenishment project overnight, will somehow bypass the area. The inevitable hurricane, moreover, is viewed as a catastrophic act of God or a sudden stroke of bad luck for which one cannot plan. The increased potential for damage resulting from shoreline engineering is also ignored in most cost evaluations. In fact, very few shoreline engineering projects would be funded at all if those controlling the purse strings realized that they must be perpetual.

"Once you begin shoreline engineering, you can't stop it!" This statement, made by a city manager of a Long Island Sound community, is confirmed by shoreline history throughout the world.

Fig. 3.9. Miami Beachless in 1972. Recently the beach was replenished over a distance of 15 miles at a cost of $65 million.

Because of the long-range damage caused to the beach it was meant to protect, this engineering must be maintained indefinitely. Its failure to allow the sandy shoreline to migrate naturally results in a steepening of the beach profile, reduced sand supply, and, therefore, accelerated erosion. Thus, once man has installed a shoreline structure, "better"—larger and more expensive—structures must subsequently be installed, only to suffer the same fate as their predecessors.

History shows us that there are two situations that may terminate shoreline engineering. First, a civilization may fail and no longer build and repair its structures. This was the case with the Romans, who built mighty seawalls. Second, a large storm may destroy a shoreline stabilization system so thoroughly that people decide to give it up. In America, however, such a storm is usually regarded as an engineering challenge and thus results in continued shoreline-stabilization projects.

Disastrous storms are not unnatural events in coastal areas. Severe storms are reasonably commonplace, and the coastal zone is the buffer that absorbs much of the storm energy. Those who live in this buffer zone should plan as though the storm of the century is coming tomorrow. It might!

A rational attitude

1. Adopt a nonstructural response whenever possible and recognize that such a response is a more ecologically sound form of protection than building physical barriers in front of your property.
2. Refuse to become a political force advocating structural responses.
3. Design to live with the flexible island environment. Don't fight nature with a "line of defense."

4. Consider all man-made structures near the shoreline temporary.
5. Accept as a last resort any engineering scheme for beach restoration or preservation, and then only for metropolitan areas.
6. Base decisions affecting island development on the welfare of the public rather than the minority of shore-front property owners.
7. Let the lighthouse, beach cottage, motel, or hot dog stand fall when its time comes, or move it out of the way.

4. Selecting a site

If you are paying a visit to the south shore, you will find that the following information will deepen your enjoyment. If, on the other hand, you plan to be a resident, it could help you keep from making costly errors in site selection.

General indicators of shoreline stability

There are a number of environmental attributes that are often good indicators of the natural history of a given area of shoreline. For the south shore the best of these are vegetation, dune profiles, dune height, terrain and elevation, and beach width.

Vegetation. Vegetation may indicate stability, age, and elevation. In general, the higher and thicker the growth of the vegetation, the more stable the site and the safer it is for development. Maritime forests like those on parts of Fire Island grow only at elevations high enough to prevent frequent overwash. In addition, since a mature maritime forest takes at least 100 years to develop, forest areas are generally the safest sites on any island.

Presence of salt-marsh grasses nearby is a sure indicator that the elevation is very low and probably subject to flooding from even modest storm surges.

Dune profiles. The profile of a dune is a good indicator of recent changes in the shoreline. If the seaward side of a dune is trimmed off along a straight line and its profile is very steep, then it has

been recently attacked by waves. If these characteristics are common along several miles of the beach, you can be reasonably sure that the beach is actively retreating. On the other hand, if the seaward face of the dune is gently sloping and covered with beach grass, the shore has been stable for at least 5 to 10 years. The toe of such dunes commonly has lines of beach grass that appear to be encroaching upon the shore.

Dune height. The height of a dune is sometimes a good indicator of longer-term stability of the beach. It takes time for dunes to grow, and the higher a dune is, the slower the growth. You must be careful in applying this criterion to the south shore, however, since areas that were once reasonably stable have recently been destabilized by the construction of jetties and groins and by the maintenance of inlets. The result is that some very high dunes now exhibit profiles characteristic of erosion.

Terrain and elevation. The terrain and elevation of an area are also measures of its safety from adverse natural processes. Low, flat areas are subject to destructive wave attack, overwash, and storm-surge flooding (see chapters 1 and 6). High-water marks up to 15 feet above sea level were produced by the 1938 storm, so even the 10-foot elevation currently required by local ordinances adopted by communities that participate in the NFIP is no guarantee that a homeowner won't experience water damage.

The position of a building relative to surrounding dunes can provide added safety or danger. A high dune south of a building has a shielding effect, but a low spot in the dune is an invitation to overwash or, worse, inlet formation. In the 1938 storm, dunes up to 18 feet high were commonly obliterated by waves.

Beach width. The width of the beach is a poor indicator of shoreline stability. The summer visitor that looks at the wide beach after a prolonged period of gentle seasonal waves is apt to jump to the conclusion that the shore is stable. The same beach may be nonexistent in late winter or early spring when storm waves arrive.

Site analysis

Montauk Point to Hither Hills (figure 4.1)

It is reasonable to begin a discussion of the south shore at Montauk Point because the gradual destruction of this scenic spot is generating the sediment necessary to maintain the beaches that lie immediately west of it. The site for the lighthouse at Montauk was chosen in 1792, and 4 years later the light was erected. George Washington is supposed to have ordered it built far enough from the edge of the bluff that it might survive the ravages of erosion for 200 years. It was built 300 feet from the bluff and now stands only about 75 feet from the edge (fig. 4.2). It is interesting that early Americans exhibited a degree of prudence in their site selection that has been lost in our current desire for an unobstructed view of the sea.

The shore from Montauk Point westward to the village of

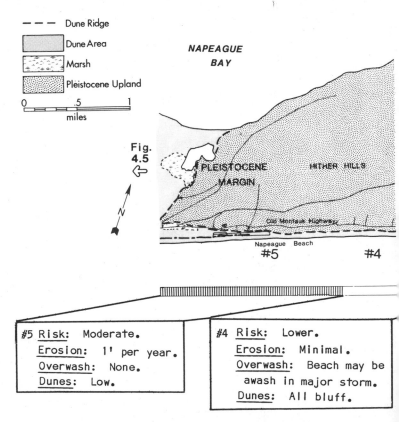

Fig. 4.1. Site analysis: Montauk Point to Napeague Beach.

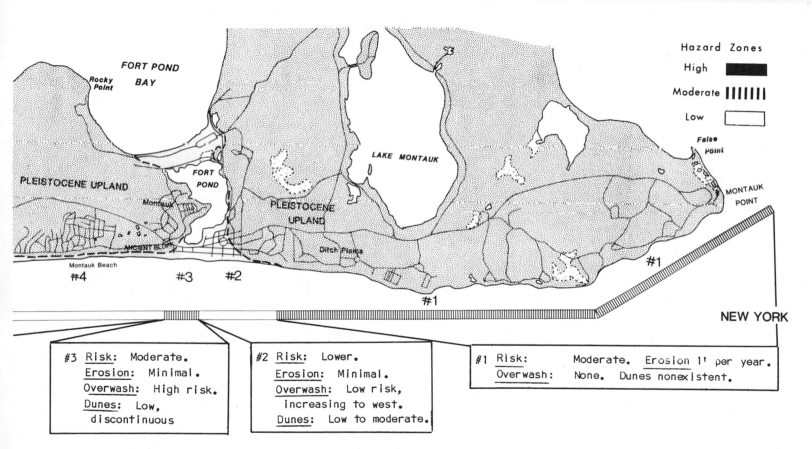

Hazard Zones

High ■

Moderate |||||||

Low □

FORT POND BAY

Rocky Point

PLEISTOCENE UPLAND

Montauk

ANCIENT BLUFF

Montauk Beach

FORT POND

PLEISTOCENE UPLAND

LAKE MONTAUK

Ditch Plains

False Point

MONTAUK POINT

#4   #3   #2   #1   #1

NEW YORK

#3 Risk: Moderate.
Erosion: Minimal.
Overwash: High risk.
Dunes: Low,
    discontinuous

#2 Risk: Lower.
Erosion: Minimal.
Overwash: Low risk,
    increasing to west.
Dunes: Low to moderate.

#1 Risk: Moderate. Erosion 1' per year.
Overwash: None. Dunes nonexistent.

Fig. 4.2. Reproduction of two undated postcards showing the dramatic retreat of the cliffs at Montauk Point. Postcards are from the collection of Charles Hutunen.

Montauk receded an average of 100 feet between the first water-line surveys in 1838 and the most recent survey in 1956; this indicates a mean erosion rate of about 1 foot per year. As explained in chapter 1, however, the retreat has been very erratic. For example, at one point the sea moved landward 300 feet, and it is probably impossible to predict exactly where such extreme rates will occur.

One of the basic rules in the study of coastal processes is that shorelines tend to straighten given enough time. Areas that jut out (promontories) are eroded and areas that indent (re-entrants) are filled with the eroded sediment. Despite this, the Montauk shore is quite irregular (fig. 4.3). The irregularity of the shoreline in this area is not explained by high and low land areas along the bluffs (which is a common reason for exceptions to the rule) but rather by the material that composes the bluffs. The glacial deposits from which the cliffs have been carved are composed of deposits that contain local concentrations of large boulders. As this material is eroded, the finer sediment is carried off by waves. The boulders simply settle into the sea to form a natural barrier that reduces the destructive force of waves. The result is that areas without this natural protection have eroded to a greater extent to produce the irregular shoreline we see today.

It would be incorrect to conclude from this discussion that bluff areas protected by natural boulder fields are retreating at a slower rate than adjacent areas. As promontories and their associated boulder fields develop, the bottom contours that are formed cause waves to focus more of their energy on the promontory than on adjacent beaches. A balance is developed between the boulders that break up wave energy and the tendency for higher wave energy to be focused on the promontories. The result is equal retreat of re-entrants and promontories so that irregularity of the shore is maintained as the entire bluff retreats (fig. 4.4).

For the prospective purchaser of bluff property this means that the promontories defended by boulder ramparts are no safer to build upon than the shallow re-entrants. Like George Washington, one should allow a generous set-back of about 300 feet.

Montauk Village is built on a low saddle in the glacial moraine (a ridge of glacial debris) that forms the backbone of the south fork of Long Island. As sea level rose to flood this low area, the westward littoral drift built a sand bar across what is now the south end of Fort Pond (fig. 4.1). Simultaneously, sand eroded from the east and west sides of Fort Pond Bay formed a second bar, enclosing Fort Pond to the north. Prior to development of these land bridges, waves carved a bluff, bordered now by Old Montauk Highway. Sediment accumulated in the embayment formed by the intersection of the bluff and the bar. As a result, the eastern end of the bluff is now separated from the sea by a relatively wide sandy strip that grows progressively narrower toward Napeague Beach. It is worthy of note that this area, which shows evidence of prehistoric seaward growth, continues to be one of the few areas on Long Island that exhibits a relatively stable water line. By contrast the short section of beach lying just east of the ancient bluff at the south end of Fort Pond is subject to flooding by storm surge, as is the bar at the north end of Fort Pond.

While the beaches and bluffs west of Fort Pond have exhibited unusual stability for a distance of 2 miles, a word of caution is appropriate. As the headland area to the east continues to retreat, Montauk beaches will be forced to recede, but that time may be a generation or two in the future.

The stability of the shore decreases westward from Montauk Beach to Hither Hills State Park. Along the Hither Hills beaches, surveys of the water line indicate a retreat of 1 to 2 feet per year. This retreat has caused some problems for permanent structures in the park camping area.

Detailed descriptions and recommendations for numbered areas in figure 4.1 follow.

Fig. 4.3. Aerial view of the irregular shoreline along the eastern headland area showing boulder ramparts in front of promontories. Montauk Village is in the background.

1. **Montauk Point to Ditch Plains**
   Risk: moderate.
   Erosion: average rate is 1 foot per year, but very erratic.
   Overwash: none.
   Dunes: nonexistent.
   Caution: construction of any permanent structure along bluffs should allow set-back of 200 feet.

2. *Risk:* lower.
   Erosion: beach recession rate is minimal.
   Overwash: no risk on east side of this area; risk increases toward west end of area.
   Dunes: low to moderate in height.
   Note: structures built above the beach sand on the Pleistocene upland are secure.

3. *Risk:* moderate.
   Erosion: beach recession is minimal.
   Overwash: high risk of overwash during major tropical storm.
   Dunes: low, discontinuous.

4. *Risk:* lower.
   Erosion: minimal in some places; historical records indicate beach has widened.
   Overwash: no risk, but entire beach may be awash in major storm.
   Dunes: none; bluff locations are reasonably secure, with risk of bluff erosion increasing gradually toward west end of area.

5. *Risk:* moderate.
   Erosion: average rate is 1 foot per year. Recent erosion threatens portions of campgrounds at Hither Hills State Park.
   Overwash: none.
   Dunes: low, increase in height to west.

Napeague Harbor to East Hampton (figure 4.5)

The Napeague Harbor area is one of the most interesting and pristine coastal areas east of Fire Island National Seashore. The land forms on this low sandy section of coast tell an eloquent story

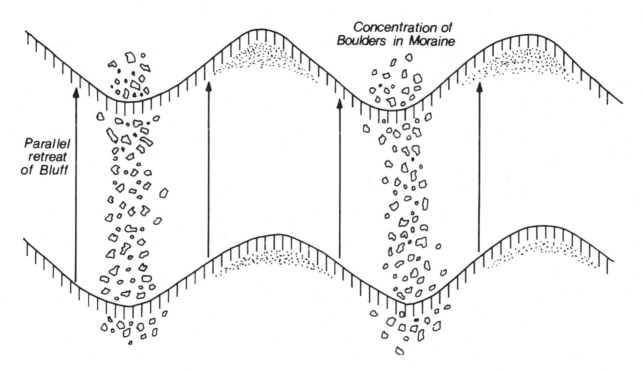

Concentration of Boulders in Moraine

Parallel retreat of Bluff

Fig. 4.4. The parallel retreat of an irregular shoreline preserves its shape.   (See figure 4.3.)

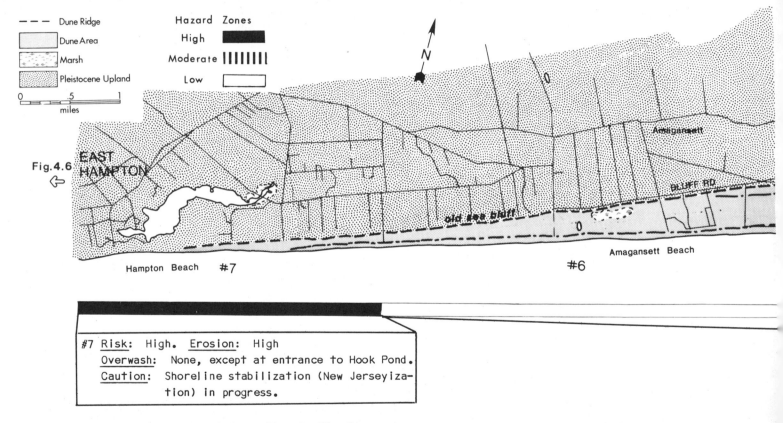

- - - Dune Ridge

Dune Area

Marsh

Pleistocene Upland

Hazard Zones

High ▬▬▬

Moderate ||||||||

Low ▭

0   .5   1
miles

Fig. 4.6

EAST HAMPTON

Amagansett

BLUFF RD.

old sea bluff

N

0

0

Hampton Beach   #7

#6   Amagansett Beach

#7 Risk: High. Erosion: High
Overwash: None, except at entrance to Hook Pond.
Caution: Shoreline stabilization (New Jerseyiza-
tion) in progress.

Fig. 4.5. Site analysis: Napeague Harbor to Hampton Beach.

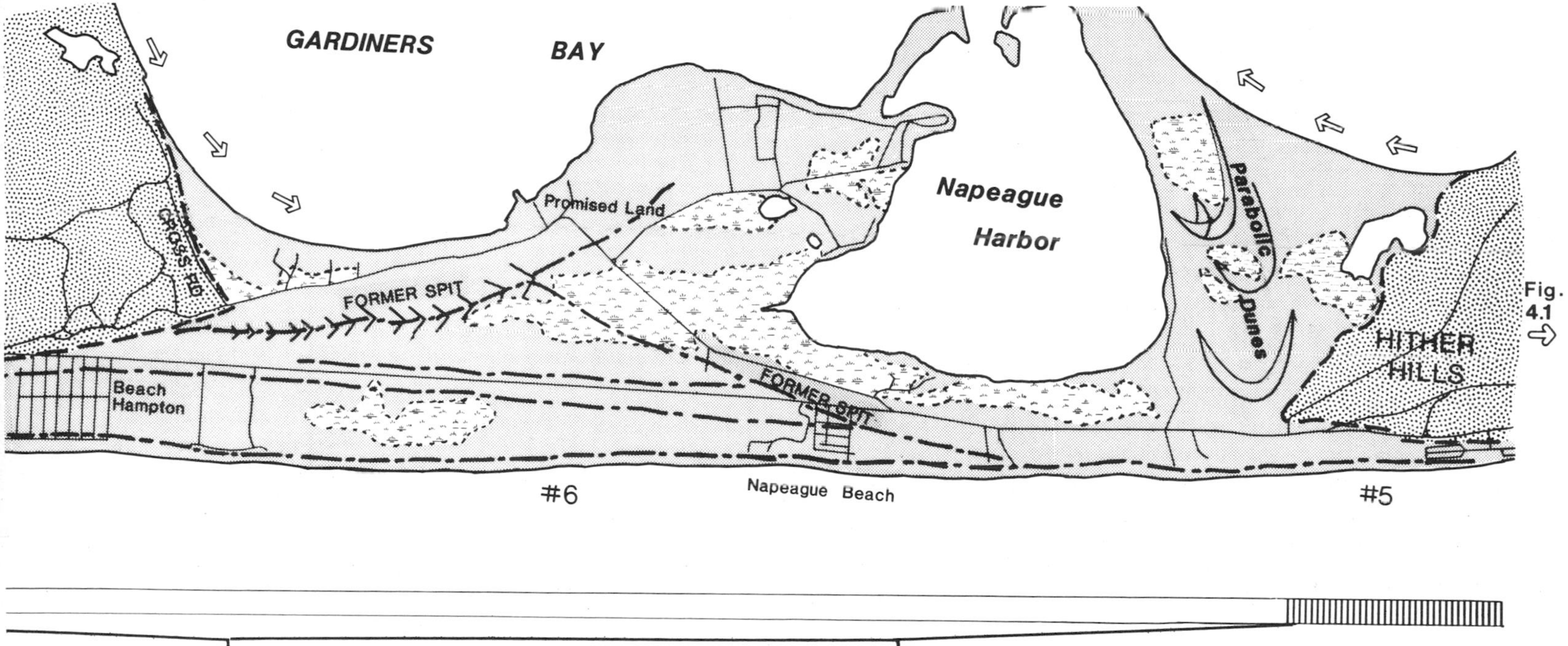

Fig. 4.1

GARDINERS BAY

Promised Land

Napeague Harbor

CROSS RD

FORMER SPIT

Beach Hampton

FORMER SPIT

Parabolic Dunes

HITHER HILLS

#6

Napeague Beach

#5

#6 Risk: Lower. Erosion: Minimal.
Overwash: Moderate risk.
Dunes: High, discontinuous.
Caution: Areas below 10' elevation will probably
flood in next major storm.

of its prehistoric formation and suggest the changes it will experience in the future. The ridge of glacial moraine forming the south fork of Long Island appears to end abruptly about 1 mile east of Amagansett Village and does not appear again until Hither Hills (figs. 4.1 and 4.5). The low, sandy land that bridges this gap has been created entirely by deposition of sand from the littoral drift of sediment eroded from bluffs to the east.

As sea level rose in response to the melting of glacial ice, this gap in the ridge must have been flooded by the sea, connecting Napeague Bay on the north with the Atlantic Ocean. At that time Hither Hills was an island, and land's end was at the intersection of Bluff Road and Cross Road. Waves carved away the glacial deposits at Hither Hills and carried the sand west to build a spit that arched toward the old Promised Land Fish Plant. A second spit built west from a small island at Promised Land toward land's end east of Amagansett. The exact position of the latter spit is indicated today by a prominent dune ridge that developed along the length of the spit. The supposition that this spit grew from east to west is supported by the height of the dunes along the ridge. They grow progressively lower to the west, as would be expected if the west end of the old spit is younger than the east end. Despite the regular decrease in dune height to the west, which agrees with the present direction of littoral drift, some scientists have questioned the direction of spit growth.

Before the two spits completed the land connection between Hither Hills and Amagansett, waves cut a sea cliff along a line that is presently bordered by Bluff Road. This ancient bluff and the two spits formed a broad, open V along the shore that trapped sediment being carried westward. A series of dune ridges is testimony to the gradual infilling of this area. The filling has progressed until the ancient spits and bluffs, which were at one time the water's edge, are separated from the ocean by as much as 3000 feet.

At the same time littoral drift on the ocean side was being trapped, erosion of material along the edges of Napeague and Gardiners Bay filled in the area north of the old spits. The result is that today the dune ridge, which marks the original location of the spits, occupies a position in the middle of the Napeague dune area. Another interesting result of infilling from the bay side is the development of three very large dunes on the eastern margin of Napeague Harbor. As the shore built northward, dunes that originated along the beach were left behind and were reshaped by the prevailing northwest winds into three large U's oriented to the southeast. The northernmost and youngest of these dunes is still quite active.

On the ocean side the prehistoric buildup that has occurred in the Napeague area appears to be continuing into the present. Surveys of the high-water line over the last 100 years indicate that the shore is either stable or building seaward. This trend must ultimately be reversed as headland areas on either side are cut back, but that reversal might be generations into the future. Even though the Napeague beaches are reasonably stable, most of the surface lies at elevations of less than 10 feet and is subject to flooding by storm surge from both bay and ocean sides. During the 1938 hurricane virtually the entire area was flooded except for the higher dunes. It was completely cut off from points to the east. This area will almost certainly see extensive flooding in the future.

Detailed descriptions and recommendations for numbered areas in figure 4.5 follow.

6. *Risk:* lower.

   Erosion: minimal; historical records show beaches widening in some areas; erosion problem worsens to west.

   Overwash: moderate risk.

   Dunes: high but discontinuous.

   Caution: this area is exceedingly low except for Pleistocene uplands. It will be subjected to extensive flooding in the next major tropical storm. All areas below 10 feet in elevation will probably flood.

7. **Hampton Beach**

   Risk: high; risk decreases toward east and west boundaries of this area.

   Erosion: moderate until mid-1970s, when beaches began to recede at a dramatic rate. This change is probably related to changes in littoral drift and presence of groins to west (see text for explanation).

   Overwash: none, except at entrance to Hook Pond.

   Dunes: dune sand mantles the eroded terminus of the outwash plain.

   Note: shoreline-hardening structures have been used exten-sively in this area, causing beaches to narrow (New Jerseyi-zation is in progress).

East Hampton to Shinnecock Inlet (figure 4.6)

This portion of the shoreline is the truncated edge of a gently sloping plain deposited by outwash from the toe of the glaciers that occupied the south fork of Long Island. Like the area to the east, it can be considered a headland but it does not exhibit the high cliffs and cobble beaches that point to an obvious erosional trend. None the less, it is an area of net sediment loss (see chapter 2) that experienced considerable prehistoric retreat. For example, a simple projection of the form of the existing outwash plain out to sea requires a minimum retreat of 1 mile between Georgica Pond and Mecox Bay (fig. 4.6). The prehistoric trend continues to the present as indicated by surveys of the water line. The average rate of retreat is about 1 foot per year in areas that have not been affected by works of man.

Shore processes and natural retreat have been strongly affected in two localities along this reach of coast. The first occurs just east of Georgica Pond in the village of East Hampton where three stone groins were built. The second is at Shinnecock Inlet where the natural inlet was widened and stabilized by the construction of stone jetties.

Although careful survey data on the effect of the East Hampton groins is not available, it appears that these structures have created an interesting and unique situation. The two largest groins were built in 1969 to stem erosion that was about to claim some expen-

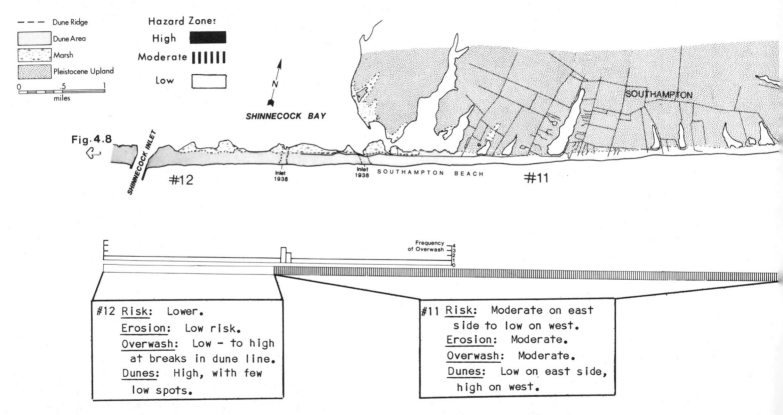

Fig. 4.6. Site analysis: East Hampton to Shinnecock Inlet.

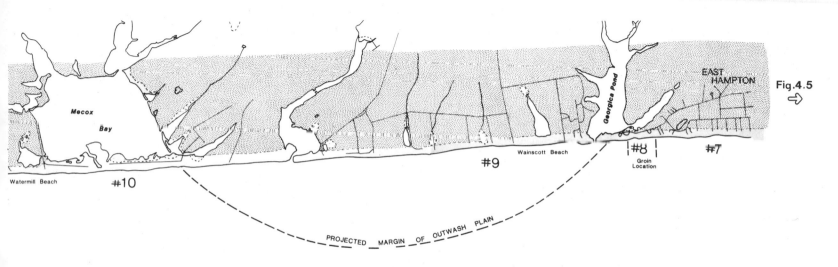

Fig.4.5 ⇨

#10 Risk: Moderate.
   Erosion: 1' per year.
   Overwash: Moderate, greatest
    at ponds.
   Dunes: Low to moderate.

#9 Risk: High to moderate. Erosion: High.
   Overwash: Low. Dunes: Low to moderate.
   Note: Most homes moved back for protection
    against erosion.

#8 Risk: Lower.
   Erosion: Moderate to low.
   Overwash: Low risk.
   Dunes: Moderate height.

sive real estate. During the next few years a home less than 1 mile west of the groins was destroyed and many other homes within 2 miles west of the groins had to be moved back from the advancing sea. The problem on the west side abated in the mid-1970s, but the homes to the east then came under severe attack. The reaction of the wealthy homeowners on the east side has been to erect a nearly continuous revetment along the shore of the village of East Hampton to harden it against further retreat. As retreat continues, the beaches in front of the revetment are disappearing (fig. 4.7).

It seems possible, even likely, that the balance point between east and west littoral drift has shifted from one side of the groins to the other. This shift probably occurs in response to a slight change in the mean annual winds that create the waves that strike this shore. The unfortunate result has been that the groins have produced negative effects on both sides.

The jetty on the east side of Shinnecock Inlet has stabilized beaches for a distance of about 1 mile to the east. The effects downdrift of the inlet have been adverse and will be considered in greater detail in the next section.

Detailed descriptions and recommendations for numbered areas in figure 4.6 follow.

8. East Hampton groin area
Risk: lower.
Erosion: moderate until installation of groins in 1966; this has significantly slowed the erosion rate.
Overwash: low risk.

Fig. 4.7. Waves awash at the toe of a bulkhead in the village of East Hampton.

Dunes: moderate height; they rest on top of eroded terminus of outwash plain.

9. Wainscott Beach
Risk: high decreasing to moderate at boundaries of area.
Erosion: prior to installation of groins in East Hampton, beaches were eroding about one foot per year. Erosion rate increased greatly after installation of groins. This change is

probably related to changes in littoral drift and presence of groins to east (see text for explanation).

Overwash: risk of overwash is slight where streets cut through dune line.

Dunes: low to moderate height.

Note: most homes along this beach have been jacked up and moved back at least once to protect them from the ravages of erosion.

10. Watermill Beach

Risk: moderate.

Erosion: beaches are eroding approximately one foot per year. This rate has been maintained over a long period of record.

Overwash: probability of overwash is greatest where ponds lie behind beach and where the ends of street cut dune line.

Dunes: moderate height with some low intervals.

Note: a number of homes in this reach have been moved back, and a number of older homes now occupy positions perilously close to the beach.

11. South Hampton Beach

Risk: transition from moderate on east side of zone to low on west side.

Erosion: until jetty was built at Shinnecock Inlet, the rate of retreat was about 1 foot per year. Erosion rate continues to be moderate on east side of this area, but impoundment of sand by the east jetty has stabilized beaches toward the west side of this zone.

Overwash: moderate risk of overwash in areas where dune line is low.

Dunes: low on east side increasing in height to west. There are several low spots in the dunes even to the west where overwash can be expected in any major storm.

Note: most of this area experienced some overwash in the 1938 storm.

12. *Risk:* lower-risk area.

Erosion: not a serious problem along this beach beacause of sand accumulation east of the jetty at Shinnecock Inlet.

Overwash: very probable overwash at low spots in the dune line; otherwise risk of overwash low.

Dunes: high dunes with a few low spots that have been the focus of recent overwash.

Shinnecock Inlet to Moriches Inlet (figure 4.8)

From Shinnecock Inlet westward the main portion of Long Island is shielded from direct wave attack by a chain of sandy barrier islands. Access to the island between Shinnecock and Moriches Inlets is via two drawbridges in Westhampton Beach— one in Quogue and one in Hampton Bays. Prior to a storm these bridges are likely to be crowded or, worse, open as both vessels and vehicles attempt to take shelter. Allow yourself plenty of time if you must evacuate this barrier segment when the summer crowd is present.

This section of the south shore has been so greatly modified by

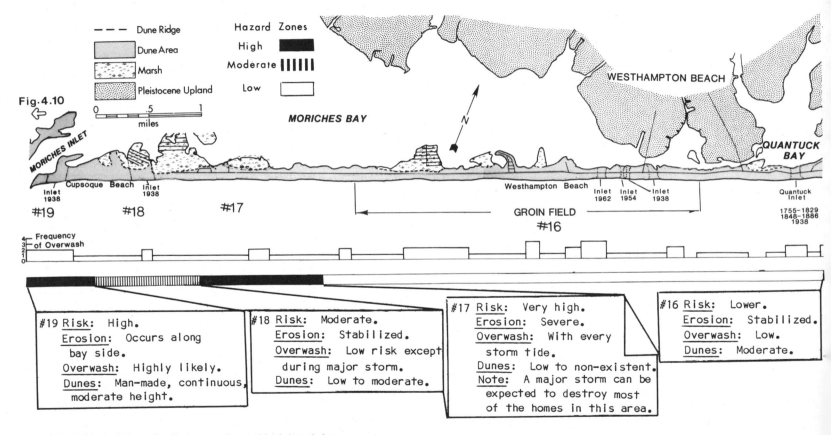

Fig. 4.8. Site analysis: Shinnecock Inlet to Moriches Inlet.

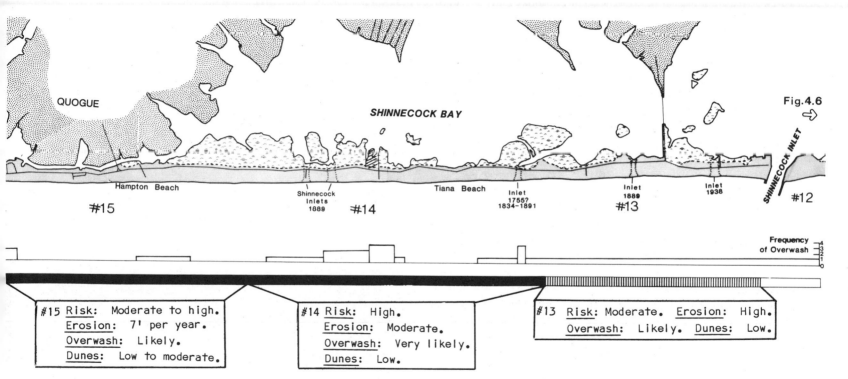

QUOGUE

SHINNECOCK BAY

Fig.4.6

Hampton Beach

Shinnecock
Inlets
1889

Tiana Beach

Inlet
1755?
1834-1891

Inlet
1889

Inlet
1938

SHINNECOCK INLET

#15

#14

#13

#12

Frequency
of Overwash

#15 Risk: Moderate to high.
Erosion: 7' per year.
Overwash: Likely.
Dunes: Low to moderate.

#14 Risk: High.
Erosion: Moderate.
Overwash: Very likely.
Dunes: Low.

#13 Risk: Moderate.   Erosion: High.
Overwash: Likely.   Dunes: Low.

works of man that historic shoreline recession rates give little indication of future trends. Between 1838 and 1930 the beach extended unbroken from Southampton to Fire Island Inlet. During this period the rate of retreat of the beach front was about 1 foot per year. Then in 1931 Moriches Inlet was opened, and 7 years later Shinnecock was opened by the great storm of 1938. Both of these inlets would have been long since closed by longshore drift except for periodic dredging and the construction of stone jetties in the 1950s. The opening and closing of inlets has been a rather common event on south-shore beaches (figs. 4.8 and 4.10). The continued maintenance of these two inlets cut off much of the natural sand supply to intervening beaches, and the rate of recession increased dramatically to an average of about 7 feet per year! The weakened shore was breached in the storm of 1962 with considerable loss of private property. The response was a cry for protection by local interests, and finally the construction of eleven stone groins in 1965–66. The predictable result was a sudden increase in erosion of the beaches immediately downdrift (west) of the last groin. To alleviate this problem four more groins were built in 1969–70 and filled with sand. Beaches within the field of groins have partially stabilized, but the area downdrift of the last groin has suffered terribly. Some homes have already been destroyed and others await their turn (fig. 4.9).

Only the gentleness of the weather has allowed the homes downdrift to survive this long. The road is awash with sand and sea in even a modest storm, and dunes do not exist. To complete this scenario the U.S. Army Corps of Engineers recently proposed

Fig. 4.9. Three homes awaiting destruction in the overwash area west of the groin field in Westhampton Beach.

to pump $42 million worth of sand onto the Westhampton beaches from the sea, but outraged taxpayers objected so strenuously to the proposal that it appears to be a dead issue (until the next storm).

At the same time that homes on the beach are threatened by loss of sand, Moriches Inlet is slowly being choked with sediment derived from the beaches. The Coast Guard has declared Moriches not navigable, but many vessels continue routine use of the inlet. The result has been a series of drownings as vessels capsize in the shoals. Some of the local fisherman now refer to it as "Killer Inlet" rather than Moriches Inlet.

While Shinnecock and Moriches inlets are recognized by most geologists and engineers to be the root cause of much of the erosion problem on nearby beaches, the solution chosen was to leave the inlets alone and build groins on the beach. These in turn have simply shifted the problem to the downdrift area. It is a classic case of understanding the technical problem but choosing a solution that only served the immediate needs of a vocal minority with vested interests. Unhappily, the problem does not stop at Moriches Inlet. Fire Island, the next link in the barrier chain, is the heir of actions taken to the east.

Detailed descriptions and recommendations for numbered areas in figure 4.8 follow.

13. *Risk:* moderate.

Erosion: rate of erosion increased dramatically after Shinnecock Inlet was widened and stabilized in 1953. Continued erosion just west of the inlet has resulted in overwash in the area of the buildings and docks there.

Overwash: likely near the present inlet.

Dunes: low. Dunes adjacent to the inlet were rebuilt artificially following overwash in the mid-1970s.

Note: expansion of dock facilities and buildings is planned for this area despite the risk of damage from overwash which is likely to occur.

14. *Risk:* high.

Erosion: moderate.

Overwash: very likely. This portion of barrier is narrow, and the bay is separated from the ocean only by low, man-made dunes. Inlets have formed repeatedly at this point.

Dunes: low.

15. Hampton Beach

Risk: moderate to high.

Erosion: rate was about 1 foot per year until Moriches and Shinnecock inlets were opened, and then the rate increased dramatically to about 7 feet per year. It is difficult to predict whether this high rate will continue.

Overwash: Likely to occur near center of the area where dunes are low and there is a past history of frequent overwash.

Dunes: low to moderate height; fairly continuous ridge.

16. West Hampton Beach

Risk: lower.

Erosion: a serious problem until groins were built. The beaches within the groin field and about one mile east have largely been stabilized by these structures.

Overwash: low probability.

Dunes: moderate height and continuous.

Note: there was a high frequency of overwash in this area prior to construction of the groins. The overwash problem has been transferred west of the last groin.

17. *Risk:* very high.

Erosion: this area has suffered dramatic and severe erosion as a result of the groins to the east that trap sand from the littoral drift and deflect the sand into deeper water.

Overwash: can be expected to occur with every storm tide.

Dunes: low to nonexistent in some parts of this area.

Note: a major storm can be expected to destroy most of the homes in this area.

18. **Cupsoque Beach**

Risk: moderate.

Erosion: sand trapped by the jetty at Moriches Inlet has stabilized most of this area.

Overwash: low probability of overwash except in a major storm.

Dunes: low to moderate height.

Caution: beaches in this area can be expected to remain stable as long as a connection with the east jetty of Moriches Inlet is maintained. Should this connection be broken, as it was in 1980, the beaches can be expected to retreat rapidly.

19. **Moriches Inlet**

Risk: high.

Erosion: here the erosion occurs along the bay side rather than the ocean side. Armor along part of the bay side has reduced erosion.

Overwash: not likely except if area beyond armor begins to erode.

Dunes: man-made; continuous; moderate height.

Fire Island: Moriches Inlet to Davis Park (figure 4.10)

The Fire Island segment of the south-shore barrier system is 32 miles long, and about 26 miles of this island were set aside by the federal government in 1964 as Fire Island National Seashore. At the east and west ends of the island respectively, Smith Point County Park and Robert Moses State Park provide access by road to large parking areas and beaches with high-density use. The philosophy that guides the development of the national seashore is different from the one that led to the construction of the adjacent parks. On the national seashore priority is given first to preserving the natural environment and second to providing for the public enjoyment of this resource. Accordingly stabilization structures are not allowed to be built upon the shore, and public use is limited to a level that is compatible with preservation of the natural environment. It is a philosophy that recognizes fully the dynamic nature of the land that forms the park, and attempts to live with this dynamism rather than fight it.

Although Fire Island supports twenty small communities, it

represents the most natural portion of the south-shore barrier island system. At first glance it seems to be a simple linear dune line adjacent to an ocean beach, but the height and position of the dunes suggests a complex history. Dunes require time to grow to heights of 50 or 60 feet. An area of continuous low dunes suggest that the beach is shifting landward so rapidly that dunes do not have time to build up. For a distance of 2.5 miles east of Moriches Inlet the dune ridge is low, rarely exceeding a height of 10 feet (fig. 4.10). This suggests that the interruption of littoral drift by the present inlet and its predecessors probably has accelerated the retreat of the shoreline in this area.

Westward to Ridge Island a single dune ridge gradually increases in height to elevations of 80 feet. Although the dunes south of Ridge Island are currently being undercut, they reflect a former period of greater stability.

Springing from a point of marsh just south of Ridge Island, a dune line arches westward until it is nearly parallel to the primary dune along the ocean. At Robinson Cove it merges with the primary dune. This line of dunes marks the western side of Old Inlet and the front of the barrier island in the mid-to-late 1700s. As the westward drift filled the inlet, the shore built seaward, straightening the coast and isolating the old dune line behind the present coastal dunes. Westward from Robinson Cove to Davis Park the dunes gradually narrow as the ridges merge. Surveys of the high-water line in 1838 and 1956 indicate that the beaches between Old Inlet and Davis Park continue to be relatively stable, but considerable erosion has recently occurred westward from Davis Park

to Fire Island Pines. Evidence of this erosion is the destruction of the primary dunes. Although data are lacking to support the contention that Moriches and Shinnecock inlets are the root cause of the reversal of the depositional trend in the western portion of this area, the idea does seem reasonable.

Detailed descriptions and recommendations for numbered areas in figure 4.10 follow.

20. *Risk:* high.
    Erosion: dune profile indicates recent deposition followed by erosion. First tier of homes is not in immediate danger, but probability is high that current erosional trend will place these structures at very high risk in the next few years.
    Overwash: likely at breaks and low spots in dune line.
    Dunes: low to moderate height; discontinuous.

21. **Eastern Davis Park**
    Risk: high.
    Notes: dune profile indicates recent deposition followed by erosion. First tier of homes is not in immediate danger, but the probability is high that the current erosional trend will place these structures at very high risk in the next few years. Second- and third-tier homes are more secure, but foundations of most homes will not withstand high-velocity currents associated with overwash.

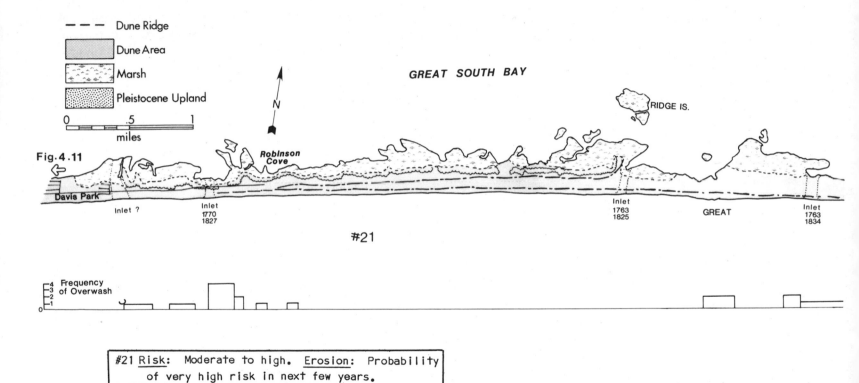

- - - Dune Ridge

Dune Area

Marsh

Pleistocene Upland

0   .5   1
miles

N

GREAT SOUTH BAY

RIDGE IS.

Fig. 4.11

Robinson Cove

Davis Park

Inlet ?

Inlet
1770
1827

Inlet
1763
1825

GREAT

Inlet
1763
1834

#21

4
3
2
1
0

Frequency
of Overwash

#21 Risk: Moderate to high. Erosion: Probability
of very high risk in next few years.

Fig. 4.10. Site analysis: Moriches Inlet to Davis Park.

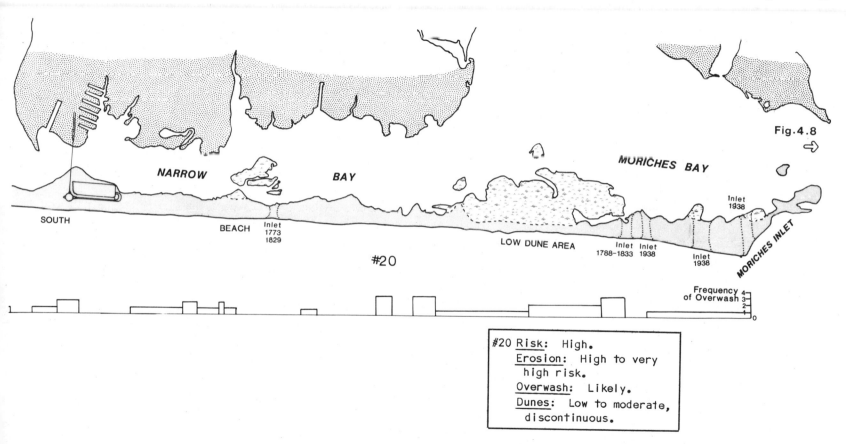

Fig. 4.8 ⇨

NARROW   BAY

MORICHES BAY

SOUTH

BEACH   Inlet
        1773
        1829

#20

LOW DUNE AREA

Inlet
1788-1833   Inlet
            1938

Inlet
1938

Inlet
1938

MORICHES INLET

Frequency 4
of Overwash 3
           2
           1
           0

1

#20 <u>Risk:</u> High.
<u>Erosion:</u> High to very
    high risk.
<u>Overwash:</u> Likely.
<u>Dunes:</u> Low to moderate,
    discontinuous.

Fire Island: Davis Park to Democrat Point (figures 4.11 and 4.12)

At Fire Island Pines the barrier island widens and the height of the dunes increases. A thicket of shrubs and trees is supported among the dunes that lie farthest from the shore.

The unusual height and width of the dunes at this locality is thought to be the result of spit growth extending the length of the barrier island.

Sand moving westward is swirled in toward the bays by flooding tides and breaking waves. The result is the construction of a spit (a small peninsula of sand) that curves gently northward into the bays (see fig. D.2). Dunes build upward on the spit and then are isolated from the sea as the spit continues to grow. If the barrier island recedes toward the land as spit growth continues, only the northward-pointing recurved tips of the spits are preserved to mark the development of the island. The west-trending dune ridge that branches from the primary dune and swings northwest to Great South Bay at Fire Island Pines appears to be such a ridge. The same holds true for the northwesterly trending ridges at Sunken Forest and Point O'Woods. Similar but less obvious ridges occur on the west side of Ocean Beach and Kismet. The lighthouse at the west end of Fire Island was built on one of these relict ridges, probably to take advantage of the natural dune height.

The distance between the lighthouse and the first clearly recognizable spit at Fire Island Pines is nearly 9 miles. The growth of that much land seems incredible until we examine the historic growth of Democrat Point. The Fire Island Light was built in 1825 only 500 feet from what was then Fire Island Inlet. It now stands 4.5 miles from the inlet (fig. 4.12). Until the jetty on the east side of the inlet was constructed, the tip of Fire Island was literally racing toward Gilgo Beach on the other side of the inlet at a rate of 200 feet per year. The jetty was completed in 1941, and the area east of the jetty was completely filled and overwhelmed with sand in less than 10 years. If we assume that the same prodigious growth occurred in the past, Point O'Woods would have formed the east side of Fire Island Inlet in 1695. In fact, historic records describe a storm in 1690 that opened the inlet to a width of 9 miles—so that Point O'Woods could easily have been the end of Fire Island. Only 75 years before that, in 1615, Fire Island Pines would have been land's end. While the dates are probably subject to considerable error, the relict sand ridges of ancient recurved spits strongly suggest that the western end of Fire Island was created in the last 300 or 400 years.

If one carries this line of reasoning to extremes, the entire 48-mile barrier system from Southampton Village to Fire Island Inlet could have been built by spit growth in only 1,200 years. Some students of the coast have advocated just such an origin. Yet these speculations bring to mind a comment made by Mark Twain in *Life on the Mississippi*. He noted that the course of the Mississippi had shortened in man's experience as river meanders were abandoned.

Now if I wanted to be one of those ponderous scientific people and "let on" to prove what had occurred in the remote past by what had occurred in a given time in the recent past, or what will occur in the far future by what has occurred in late years, what an opportunity is here! Geology never had such a chance, nor such exact data to argue from! . . . Please observe: In the space of 176 years the Lower Mississippi has shortened itself 242 miles. That is an average of a trifle over one mile and a third per year. Therefore any calm person who is not blind or idiotic, can see that in the Old Oolitic Silurian Period, just a million years ago next November, the lower Mississippi River was upwards of 1,300,000 miles long and stuck out over the Gulf of Mexico like a fishing rod. And by the same token any person can see that 742 years from now the Lower Mississippi will be only a mile and three-quarters long. . . . There is something fascinating about science. One gets such wholesome returns of conjecture out of such a trifling investment of fact.

Twain's comments seem particularly apropos regarding spit growth. There is very good evidence that indicates the barrier system is much older than the 1,000 or 2,000 years this process would require. Peat deposits were found seaward of Fire Island and dated at 7,500 years old by use of radioactive carbon. The peat must have been deposited in a bay behind a barrier that existed at least 7,000 or 8,000 years ago at a more seaward location. So, a more likely alternative to the spit-growth explanation is that the barrier islands have been in existence for a long time and have been shifting landward in response to the rising level of the sea. In any event, spit growth has undoubtedly played an important role in progressively linking and modifying island elements as the barrier chain evolved.

Detailed descriptions and recommendations for numbered areas in figures 4.11 and 4.12 follow.

22. **Western Davis Park**
    Risk: high.
    Note: first tier of homes is at very high risk. They will likely suffer damage from storms of moderate intensity.

23. **Eastern Fire Island Pines**
    Risk: moderate.
    Notes: dune profile indicates a recent advance of the dune face followed by erosion. First tier of homes is at moderate risk. Home sites behind first tier are at lowest risk on Fire Island. Many homes have flood-resistant foundations. Some homes on bay side are built below the 5-foot contour and are subject to flooding from moderate storm surges.

24. **Western Fire Island Pines**
    Risk: moderate to high.
    Notes: primary dune is increasingly truncated to west. Accordingly the risk of damage to first tier of homes increases progressively to the west. The truncation of the dune is severe, and homes are in immediate peril on the west 150 yards of this area.

25. **Cherry Grove**
    Risk: moderate to high.
    Notes: the primary dune is in the best condition on the east

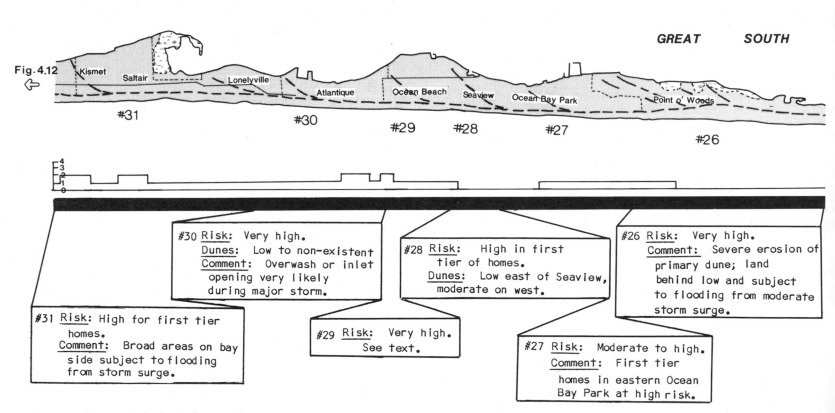

GREAT   SOUTH

Fig. 4.12

Kismet
Saltair   Lonelyville
                              Atlantique   Ocean Beach   Seaview   Ocean Bay Park   Point o' Woods

#31   #30   #29   #28   #27   #26

#30 Risk: Very high.
Dunes: Low to non-existent
Comment: Overwash or inlet
opening very likely
during major storm.

#28 Risk: High in first
tier of homes.
Dunes: Low east of Seaview,
moderate on west.

#26 Risk: Very high.
Comment: Severe erosion of
primary dune; land
behind low and subject
to flooding from moderate
storm surge.

#31 Risk: High for first tier
homes.
Comment: Broad areas on bay
side subject to flooding
from storm surge.

#29 Risk: Very high.
See text.

#27 Risk: Moderate to high.
Comment: First tier
homes in eastern Ocean
Bay Park at high risk.

Fig. 4.11. Site analysis: Davis Park to Kismet.

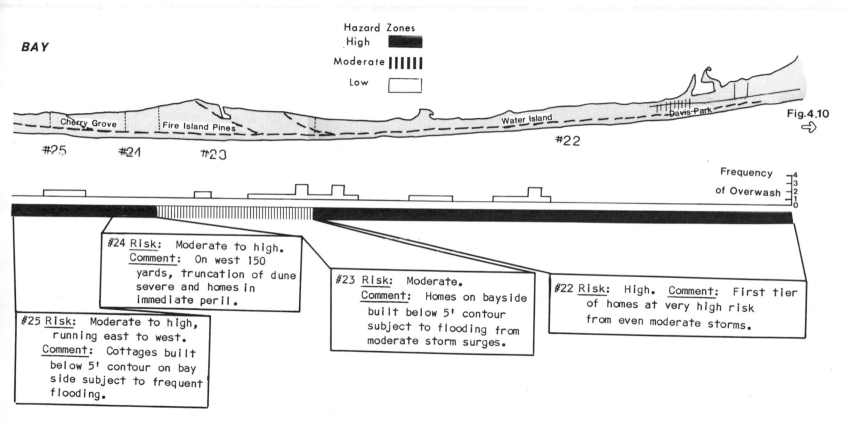

BAY

Hazard Zones
High ▐███▌
Moderate ▐||||||▌
Low ▐ ▌

Cherry Grove   Fire Island Pines   Water Island   Davis Park   Fig. 4.10

#25   #24   #23   #22

Frequency
of Overwash

4
3
2
1
0

#24 Risk: Moderate to high.
Comment: On west 150
yards, truncation of dune
severe and homes in
immediate peril.

#25 Risk: Moderate to high,
running east to west.
Comment: Cottages built
below 5' contour on bay
side subject to frequent
flooding.

#23 Risk: Moderate.
Comment: Homes on bayside
built below 5' contour
subject to flooding from
moderate storm surges.

#22 Risk: High. Comment: First tier
of homes at very high risk
from even moderate storms.

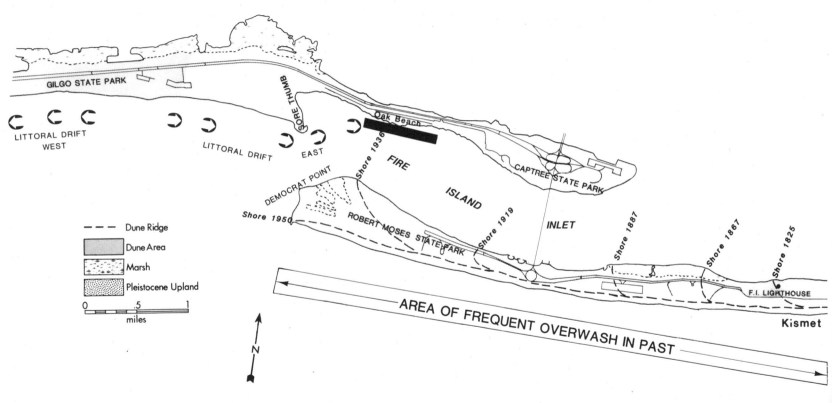

GILGO STATE PARK

LITTORAL DRIFT
WEST

LITTORAL DRIFT   EAST

SORE THUMB

Oak Beach

Shore 1936

FIRE

ISLAND

INLET

CAPTREE STATE PARK

DEMOCRAT POINT

Shore 1950

ROBERT MOSES STATE PARK

Shore 1919

Shore 1887

Shore 1867

Shore 1825

F.I. LIGHTHOUSE

Kismet

AREA OF FREQUENT OVERWASH IN PAST

- - - Dune Ridge

Dune Area

Marsh

Pleistocene Upland

0   .5   1
         miles

N

Fig. 4.12. Site analysis: Kismet to Gilgo State Park.

side of Cherry Grove and grows progressively lower to the west. Consequently, first-tier homes are at moderate risk on the east side of the village and at high risk to the west. Many cottages are built below the 5-foot contour on the bay side and are subject to frequent flooding. Home sites between the bay and the ocean are in the safest position but most have been built without flood-resistant foundations.

26. Point O'Woods
Risk: high.

Notes: primary dune has experienced severe erosion. Only a remnant of the original dune is left. Several homes occupy positions right on the beach and will be destroyed in the near future by storms of even moderate intensity. Much of the land behind the primary dune is low, making it subject to flooding from storm surge.

27. Ocean Bay Park
Risk: moderate to high.

Note: in eastern Ocean Bay Park the first tier of homes is built directly on top of the primary dune and is at high risk.

28. Seaview
Risk: high in the first tier.

Notes: homes on the east side of Seaview are built behind the primary dune, but the dune is so low that they are at moderate to high risk. On the west side the dune is in slightly better condition, but some homes are perched directly on top of it and are at high risk.

29. Ocean Beach
Risk: high in the first tier.

Notes: the first tier of homes is on top of the dune. At the west end the second tier projects beyond the first tier, and several homes are in immediate danger of destruction. Homes in the third or fourth tiers are distant enough from the beach to be safe from erosion and high enough to avoid most flooding. Many homes on the bay side are subject to flood from storm surge.

30. Atlantique and Robbins Rest
Risk: high.

Notes: dune line is low to nonexistent, making this area a very likely place for extensive overwash or inlet opening during a major storm. The danger of flooding and damage from high-velocity currents is particularly high at Robbins Rest.

31. Fair Harbor, Saltaire, Kismet
Risk: high for first tier of homes.

Notes: first-tier homes are located on top of or in front of the dune. Home sites in the third or fourth tiers are at lower risk. Broad areas on the bay side are subject to flooding from storm surge.

Comments on Fire Island communities

Fire Island supports twenty small communities that with a few exceptions are likely to suffer severe damage in time of a major storm. The areas safest from damage occur in places where the high dunes of an old recurved spits angle northwest through the center of the barrier. Fire Island Pines is most notable in this respect.

Access to the villages is gained by four-wheel-drive vehicles or passenger ferries originating at Bay Shore, Sayville, and Patchogue. This provides homeowners with a feeling of separateness from the busy roads of Long Island, and many find this attractive. At the same time, the physical separation of these communities from the mainland and the high likelihood of severe overwash in a storm increases the need to allow adequate evacuation time when a hurricane threatens. There is little question that some of those who decide to ride out the storm will be making their last mistake.

With Fire Island's summer population at 20,000 people, the evacuation is not likely to be smooth or complete. For those who either cannot or will not leave, these communities are potential death traps.

You don't need to wait for a hurricane to encounter hazards on a barrier island. The water supply for most of the homes on Fire Island comes from a few deep-water wells. Waste disposal in cesspools has long since made the water in the shallow sands unfit to drink. If you are in a high-density housing area on Fire Island you should not drink the water from shallow wells.

The communities of Fire Island Pines and Cherry Grove are protected by a primary dune line that is reasonably high; this line is not present in communities farther to the west. In both villages the primary dune is highest on the east end and decreases in height progressively to the west. Ocean-front homes on the west side of both communities are in more peril than those on the east. In addition, a number of homes in Cherry Grove are built on the crest of the primary dune. The western half of Fire Island Pines is composed of high dunes that extend all the way to the bay side of the island, providing an unusual number of home sites at high elevations and safe distances from the beach. Homes in both communities have been located without regard to elevation of the land. Many homes on the bay side are constructed less than 5 feet above high tide and are subject to frequent flooding.

Point O'Woods, Ocean Bay Park, Seaview, and Ocean Beach are the next contiguous group of communities to the west (fig. 4.11). With few exceptions the barrier island in these areas consists of a single dune ridge that grades gently back toward the bay. The dune has experienced recent erosion (fig. 4.13), and many ocean-front homes are in peril. Bay-side homes are low and subject to flooding from storm surge. Between these two extremes is a narrow strip of land, which at elevations greater than 5 feet is comparatively safe from the ravages of erosion.

Robbins Rest, Atlantique, and a small slice of the national seashore lie further west. The primary dune along this section of beach is low to nonexistent. It is an area very likely to be completely overwashed in a major storm and it is a candidate for a new inlet. This possibility is enhanced by streets oriented perpendicular to shore; these streets will provide an avenue for flood waters as well as vehicles (fig. 4.14). It is a very high-risk area.

Next come Lonelyville, Fair Harbour, Saltaire, and Kismet. The

Fig. 4.13. Homes damaged at Ocean Beach on Fire Island in 1962 storm. Source: U.S. Army Corps of Engineers.

dune line here is low and has been subject to recent retreat. Many homes project beyond the edge of the primary dune onto the beach and will almost certainly be damaged or destroyed in the near future. Most of the surface lies at elevations less than 5 feet and is subject to storm-surge flooding. Between the beach front and the lowland on the bay side is a narrow strip of land less subject to flooding and erosion.

In general, erosion problems worsen progressively to the west

Fig. 4.14. View north across Fire Island at Robbins Rest. Dunes are low or absent, and the sandy lane provides an excellent avenue for overwash.

from Fire Island Pines. The height of the primary dune decreases, and the proportion of lowland on the bay side increases.

Fire Island Inlet to Jones Inlet

The barrier island from Fire Island Inlet to Jones Inlet is entirely within the public domain. The philosophy that has guided the development of this shore is the antithesis of the national seashore policy of preservation of the natural system and non-intervention. It has been developed as a suburban recreational area with public access and human use as the first priority. Preservation of the natural barrier environments has played a subservient role.

The marshes and shoals on the bay side of the island have been buried to provide parking space, or mined to supply sediment to the beach face and dune line. In 1927 the natural dunes were buried and bulldozed with the addition of thousands of yards of hydraulic fill to form a ridge of uniform height at 14 feet above sea level. The ridge was paved to provide a broad parkway that extends from Captree Island on the east to Jones Beach State Park on the west.

Today as you drive along the Ocean Parkway you must remind yourself that this area was once as wild and natural as present-day Fire Island. It exhibits the smooth engineered appearance of an interstate highway. The creation of Ocean Parkway, parking lots, and public bathing facilities is a tangible expression of the state's commitment to arrest the dynamic nature of this barrier island. The manipulation of the shore has had mixed effects.

Captree Island, Oak Beach, and Cedar Beach form the eastern end of this barrier (fig. 4.12). Oak Beach has had a long history of inlet openings and erosion. In light of its history, a more unlikely place to build a development is difficult to imagine. Despite this, Oak Beach is one of the few places where private citizens were allowed to build homes.

In 1834 the end of Fire Island had not yet overlapped Oak Beach, and an inlet existed between Captree Island and Oak Island. The inlet later filled, but breaches and overwash continued to occur in this area. As Democrat Point built west and north, it kept the currents of highest velocity pressed against Oak Beach; this resulted in further erosion. Even though Oak Beach has received artificial replenishment, only temporary relief has been provided. Small groins and bulkheads were built, but these have proven to have only limited value. In 1959 a sand dike was built south across the offending channel to reduce the scour at Oak Beach (fig. 4.15). This dike, appropriately called "sore thumb," may have helped to reduce the scour, but it also cut off the supply of sand to Oak Beach, which further aggravates the erosion problem.

Democrat Point shields Oak Beach from waves that would produce westerly drift, while waves that cause drift to the east can pass through Fire Island Inlet and strike the shore. Consequently the littoral drift for Oak Beach is dominantly to the east. In addition tidal currents running into the inlet are concentrated on the Oak Beach side. When the sand dike was built, it stopped the sand that would have found its way to Oak Beach.

Some of the Oak Beach residents have expressed concern that the 1977 dredging of the shoals at the entrance to Fire Island Inlet has allowed waves from the open ocean greater access to their beach, and has thus caused more erosion. While this might be true, local fisherman returning through the inlet in heavy seas are likely to take a dim view of allowing the inlet to shoal so that real estate can be preserved.

Local interests have recently begun armoring the shore with slabs of waste concrete and other debris. While the effectiveness of this procedure remains to be seen, the aesthetic impact is instantaneous (fig. 4.16). Anyone considering a residence in the Oak Beach area is probably well advised to consider structures in the second tier of lots from the inlet in the expectation of being on the waterfront in the not-too-distant future.

Gilgo, Tobay, and Jones beaches to the west were originally separated by inlets, but the last of these was filled in 1930. The shore at Gilgo and Tobay has exhibited nearly continuous erosion from the time of the first survey in 1834 until the present. The recession of the beaches was worsened by the construction of the jetty at the end of Democrat Point. While the jetty eliminated the shoaling problem in Fire Island Inlet for about 9 years, it was equally successful in starving the downdrift beaches. To alleviate the problem, sand was mined from the bay to restore the ocean-side beaches. So much sand was mined that the Park Commission became alarmed that the nearby sand supply was being depleted and that the quantity remaining was not sufficient for long-range beach nourishment. After the Fire Island jetty was filled, sand settling in the inlet was dredged and placed on Gilgo Beach. Although this sand bypass operation was originally planned to occur on a periodic basis, it has been performed only as needed.

Like Fire Island, the east jetty at Jones Inlet has caused Jones Beach to build outward, so today the parking lots and bathing facilities are protected by wide beaches.

Point Lookout to Coney Island

Each link in the Long Island barrier represents a markedly different management philosophy. With the exceptions of the public beaches (Lido and Hempstead Park), the policy at the eastern end has been one of unrestrained development. Somewhat differently, the policy from Captree to Jones Beach has been for public recreation, maximizing access, and minimizing preservation of the natural barrier environments. Both of these policies are in sharp contrast with Fire Island where the national seashore is attempting to preserve the barrier in a natural state and still provide public access. From Long Beach west to the tip of Coney Island, unrestrained development has buried marsh, dunes, and beach beneath a layer of pavement and buildings.

The city of Long Beach is a remarkable example of the speed with which a natural barrier can be obliterated by development. It has been converted to an urban center as though its underpinnings were as solid and immutable as the rock foundation of Manhattan. The water front is studded with groins and lined with a wide boardwalk where dunes would normally grow. High-rise apartment buildings buttress the boardwalk (fig. 4.17), and paved

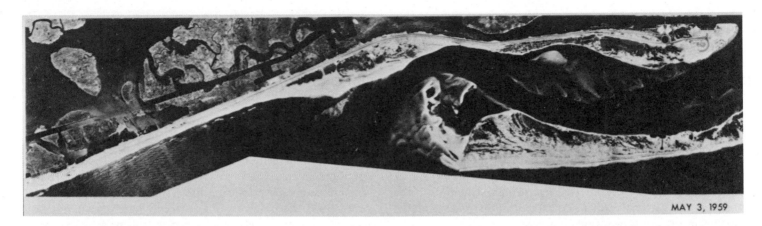

MAY 3, 1959

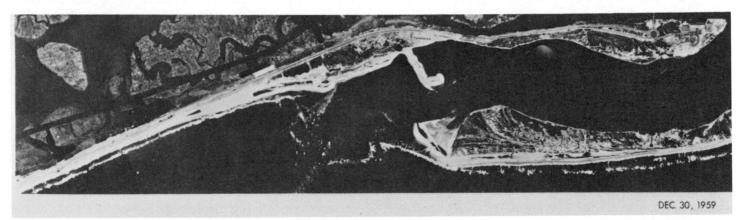

DEC. 30, 1959

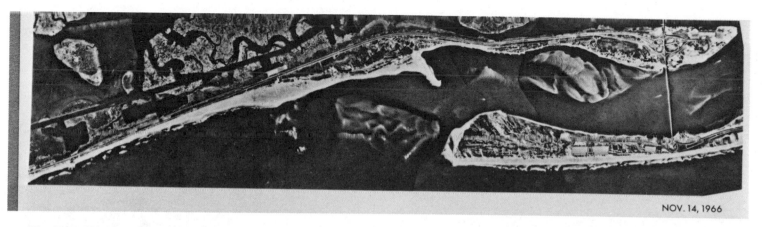

NOV. 14, 1966

Fig. 4.15. (A) Aerial view of Democrat Point and Oak Beach prior to the construction of the sore thumb dike. (B) Sore thumb dike completed and channel dredged. (C) Sore thumb accumulates sediment on west side, depriving Oak Beach to the east of its sand supply. Source: Richard J. Hannington, Department of Environmental Control, Town of Babylon.

streets extend to a commerical core that gives the feeling of downtown anywhere U.S.A.

At the time of the Civil War, Long Beach was a relatively natural barrier system. By the late 1800s it was a community of beach cottages and one large hotel. Then in 1906 a syndicate of wealthy developers purchased Long Beach, and in the winter of 1908–1909, steam dredges were brought in to pump marsh and bay sediment onto the property as a foundation for construction. The dredging operation was regarded as the biggest ever undertaken with the exception of the Panama Canal, and the amount of lumber moved was so great that elephants were employed in the task. Only 14 years passed between the time the first steam dredge spread muck over the marshes until the incorporated city of Long Beach with 25,000 residents stood on the shore. Today the population is over 260,000 and the only indication of the former barrier is the proximity of the city to the water's edge.

The surface at Long Beach was graded from a maximum elevation of 10 feet on the ocean side to only 4 feet on the bay side. The

Fig. 4.16. Slabs of concrete piled on the shore at Oak Beach to slow erosion.

Fig. 4.17. The urbanized shore at Long Beach.

entire city has its foundations below the 100-year storm-surge level, so it periodically experiences flooding. Storms in 1960 and 1962 caused about $8 million in damages, but the potential for much greater damage exists if the city is struck by a severe hurricane. Erosion along the shore persists even though a groin field has been erected along most of the beach. Long Beach is a city committed to stopping any further encroachment of the sea, and willing to accept the specter of periodic flooding.

The barrier island chain is completed with the Rockaways and Coney Island. Both have been completely urbanized (fig. 4.18), with the Rockaways experiencing a substantial amount of urban decay. The shore is studded with groins, but erosion continues to be a major problem. In the early 1970s the erosion problem at Rockaway Beach became so severe that the U.S. Army Corps of Engineers was asked to completely rebuild the shore. They did so

Fig. 4.18. A view north across Coney Island. At one time this was a barrier island separated from the mainland by a salt marsh. Now the only indication of its island origin is its name and proximity to the sea.

by dredging holes seaward of the island and moving the sand onto the beaches at the 1976—77 price of $15 million. The existing groins buried by the project have emerged once again as the sand drifts away. The original plan was to include additional replenishment when needed for a period of 10 years. What happens at the end of that 10-year period? The answer, of course, will be to dredge additional holes in the sea floor at greater and greater cost. The preservation of the city and the exceedingly heavy use of public beaches dictates a policy of eternal beach replenishment for the Rockaways and Coney Island.

The Corps of Engineers concluded from the Rockaway experience that the sea floor is a practical source of sand for the south shore. Mining of the nearshore bottom is planned in spite of evidence that south-shore beaches derive some of their sand from sources near the shore. Little consideration has been given to the effect of shore-parallel trenching on the natural onshore movement of sand. The question arises whether we might not be creating a greater problem in the future by present, well-intentioned sand mining. The common pattern of adverse environmental reactions to our attempts to manipulate the environment suggests that nearshore sand sources should be approached with the greatest care.

5. Land use and the law

The development of the New York shoreline has been haphazard. Urbanized islands adjacent to the city must now be maintained at public expense. Barriers developed for public recreation (Jones Beach to Oak Beach) have permanent structures that can only be preserved by continuous beach replenishment, a costly process. Along privately owned portions of the eastern beaches, structural solutions like those at East Hampton and Westhampton Beach have been effective only at transferring erosion problems to adjacent property. The Westhampton Beach project is an outstanding example of the effect of public pressure causing local government, and ultimately the U.S. Army Corps of Engineers, to proceed in opposition to their own guidelines.

While most of us find added regulation repugnant, it is obvious that only by increasing control of land use at the regional level can our beaches be saved from following the New Jersey example—and then only if the philosophy used in developing guidelines for control recognizes the dynamic nature of the shore. We must all be willing to live with natural changes in the shore rather than adopt the engineering philosophy of man against the sea.

The following is a brief discussion of the laws and regulations that are currently in effect for the New York coast.

The New York State Coastal Management Program

In 1972 the federal government passed the Coastal Zone Management Act which encourages coastal states to develop their own coastal management programs. For those states that create acceptable programs, the federal government provides annual funding to defray the cost of operations. In September 1982 New York received approval of its program.

The purpose of the New York State program is to provide for better utilization of the coast by balancing the need to conserve the state's coastal resources with the continued pressure for coastal development. It is hoped that these ends will be served by addressing eleven areas of concern: (1) aesthetics; (2) agriculture; (3) air quality; (4) economic activity; (5) energy; (6) fish and wildlife; (7) flooding and erosion; (8) offshore oil development; (9) coastal access by the public; (10) recreation; and (11) water quality.

To address these concerns the state has developed forty-four specific policies that are enforced using existing laws such as the federal Tidal Wetlands Act and Fresh Water Wetlands Act. In 1981 the state's Erosion Hazard Areas Act was passed.

Unlike the wetlands acts, the Erosion Hazard Act attempts to introduce control over development above the tide line in coastal areas. It will directly affect all those that plan to build in areas where erosion is a problem (most of the south shore). The state is now in the process of preparing maps of the coastal areas where this law will apply. Once completed anyone planning to build within an erosion hazard area will be required to obtain an erosion area permit and conform to specific set-back requirements. This applies to erosion protection structures such as bulkheads and revetments as well as new homes. The New York State Department of Environmental Conservation is the agency responsible for administration of the new act.

In fact, the county might issue the permit if local regulations meet state requirements. At this point questions regarding permit applications and requirements are best addressed to the New York Department of Environmental Conservation.

Importantly, this law gives the state or local government authority over erosion-control structures. Unfortunately, the position of the state regarding the use of structures to control erosion is unclear. The act states, "Construction of erosion protection structures is expensive, often only partially effective over time, and may even be harmful to adjacent or nearby property. In some areas of the coastline, major erosion protection structures of great length would be required to effectively reduce future damages due to erosion." The act goes on to say that where destruction to man-made property is likely to occur, structures may be allowed. It then requires the structure to be effective for a period of at least 30 years, assuring that if construction occurs it will be on a massive scale.

It is interesting to note that the Regional Marine Resources Council, a committee of the Nassau-Suffolk Regional Planning Board, recommended the prohibition of structural solutions unless it could be specifically demonstrated that there would be no adverse effects. The Regional Planning Board, seemingly bowing to local pressure, took a middle-of-the-road position, suggesting that structural solutions would be satisfactory along the urban coast and at Westhampton Beach. The state, by comparison, exempts no area from possible structural solutions and sets up a framework that might increase the chances of such solutions being used.

Aside from this issue, the state coastal management program attempts to move local government toward assuming a major role in developing a management program that will conform to the generally laudable goals set up by the state.

For more information contact the individuals listed in appendix B under Coastal Zone Management Program.

National Flood Insurance Program

Prior to establishment of the National Flood Insurance Program (NFIP), storm and flood victims looked to the federal government strictly for postdisaster relief. The general public shared the financial burden for those who chose to live in high-risk areas. Even though the general public still bears a considerable portion of the financial burden, the flood insurance program places a greater share of this burden directly on those that benefit. It provides insurance to homeowners who would not have been covered by private insurance companies because of the high risk. The premiums paid by owners of existing residences are subsidized by the government so that rates remain low even though the risk is high. New construction is charged premiums that are actuarial— that is, that reflect the true risk to the structure. Early in the program these actuarial rates were frequently set too low, but they are now being adjusted upward to place the burden on those that choose to take the risks. Unfortunately, storm and flood victims still look to the federal government for other forms of postdisaster relief. Likewise, insurance has not eliminated the requests for public funding or protective shore structures.

Flood insurance is available to you through local insurance agents if the community in which you live has joined the program. All of the towns along the south shore and New York City are currently enrolled. To do so, they were required to adopt a flood-plain management program that is designed to protect new construction from future flood damage.

Communities that join can be in the emergency or regular program. The emergency program imposes minimal management requirements on the local community and entitles homeowners to receive what the Federal Emergency Management Agency (FEMA) calls basic coverage at subsidized rates. FEMA then conducts an engineering study and prepares a map of the community that includes flood elevations and designates areas with various levels of risk (Flood Insurance Rate Maps or FIRMs).

The community is required to adopt a more comprehensive management plan using these data. At this point the community becomes a member of the regular program, and additional coverage can be purchased. Under the regular program, actuarial (true risk) rates are charged.

Homes built before the program went into effect can be insured at the same rate regardless of their elevation. Once the FIRM is in effect all structures must be built above the 100-year flood level (this is the elevation that flood waters have a 1 percent chance of reaching in any given year); this must be done so that structures comply with local ordinances and qualify for low insurance rates. Since October 1, 1981, this "Base Flood Elevation" includes wave heights in areas designated as V-Zones. (V-Zones are areas likely to be penetrated by waves on top of the flood level.) Before buying or building on the shore you should ask:

1. Am I in a flood hazard area?
2. Is the building above the 100-year flood level (including calculated wave height in the V-Zone)?
3. What are the structural requirements for new construction in order to comply with local ordinances and qualify for the lowest insurance rates? (Currently, there are no structural requirements for existing buildings.)
4. What are the premiums and limits of coverage?

Structural requirements vary among different towns and villages, so the best advice is to consult the local municipality for details. Local insurance agents will have the necessary information to tell you if you are in a flood hazard area and to quote premiums based on up-to-date flood insurance rate maps. Additional information regarding details of this plan is available by calling or writing a FEMA office (see appendix B under Insurance).

The federal flood insurance program has some flaws. Probably the most frequent complaint is that the boundaries of the flood insurance rate maps are inaccurate or difficult to use, but there is an appeals system designed to solve these problems. Another claim is that flood insurance encourages individuals to buy or build homes in disaster-prone areas. Studies of insurance purchasers are mixed on this score, but availability of flood insurance certainly does nothing to discourage construction in high-risk areas. In fact, along the south shore of Massachusetts several homes have been destroyed and rebuilt as many as three times, all with the aid of federal flood insurance. Prospective homeowners should be aware of ongoing changes in the flood insurance program. Already a much greater share of the cost for the program has been shifted from the tax-payer to the property owner, through a dramatic rise in premiums. This trend will continue.

Some flood insurance facts

Know the difference between a homeowner's policy and a flood insurance policy. (1) Flood insurance offers the potential flood victim a less expensive and broader form of protection than would be available through a postdisaster loan. (2) Flood insurance is a separate policy from homeowner's insurance. The latter only covers damage from wind or wind-driven rain. (3) Flood insurance covers losses resulting from the general and temporary flooding of normally dry land, the overflow of inland or tidal water, and the unusual and rapid accumulation of runoff of surface water from any source.

Check if your property location has been identified as flood-prone on the federal Flood Insurance Rate Map or Flood Hazard Boundary Map (FHBM) for an emergency program community. If you are located in a flood-prone area, you must purchase flood insurance to be eligible for all forms of federal or federally related financial, building, or acquisition assistance, including VA and FHA mortgages, SBA loans, and similar programs. To locate your property on the FIRM or FHBM, see your insurance agent. Also, keep the following points in mind: (1) You need a separate policy for each structure. (2) If you own the building, you can insure structure and contents, or contents only, or structure only. (3) If you rent the building, you need only insure the contents. A separate policy is required to insure the property of each tenant.

A condominium unit that is a traditional townhouse or rowhouse is considered for flood-insurance purposes as a single-family dwelling. The individual units may be separately insured.

Most mobile homes are eligible for coverage if they are on foundations. It makes no difference whether or not a foundation is permanent and whether the wheels are removed either at the time of purchase or while the home is on the foundation.

The following structures are not eligible for flood insurance: (1) travel-trailers and campers; (2) new mobile homes in V-Zones;

(3) fences, retaining walls, seawalls, septic tanks, and outdoor swimming pools; (4) gas and liquid storage tanks, wharves, piers, bulkheads, growing crops, shrubbery, land, livestock, roads, or motor vehicles.

One insurance broker cannot charge you more than another for the same flood-insurance policy. Rates are set by the federal government.

There is a 5-day waiting period. Coverage will not be effective until 5 days after the date of application, except when there is a transfer of title.

Building permits

Building permits for any structure to be located in a community that is part of the regular National Flood Insurance Program will require you to meet certain minimum structural requirements set by the federal government for floodproofing. The office to contact for guidance in the acquisition of the necessary building permits and regulations is the Department of Regulatory Affairs in the New York State Department of Environmental Conservation (see appendix B, under Building permits). Keep in mind that in almost all cases, local codes represent a minimum standard for protection of your investment, and you might do well to insist that some additional safety measures be built in (see chapter 6). Another point worth keeping in mind is that even a "floodproofed" home can be flooded or destroyed.

As the New York State Coastal Zone Management Program makes progress in defining erosion hazard areas, it is possible that regulations will become even more stringent for shore-front property. It seems probable that these zones will be defined within a few years of the writing of this book, so you might wish to check with the state coastal zone management program before you purchase a given piece of property.

Water supply and sewage system permits

You will need permits for water supply and sewage systems. Most of the land left available for development on the south shore lies in Suffolk County, and many of these sites will require a waste disposal system. The Suffolk County Department of Health Services has available several articles explaining the permit procedure and regulations (see appendix B under Sewage and waste and Water resources). You may also require a private water well depending on the exact location of the building site. The regulations the county imposes are based on the way in which waste fluids and water supply systems interact beneath the surface. At a shallow depth (5 to 30 feet) the pore space in the soil is completely filled with water that has filtered down from above. The boundary between the soil saturated with water and the zone above is termed the *water table*. If you have your own shallow well, you pump water from a pipe driven deep enough to penetrate the water table. You use it in your home for myriad purposes and then return it directly to the water table through your waste disposal system—it is returned complete with a broad spectrum of pollutants. If wells and disposal beds are located too close together, the water near the well becomes increasingly polluted.

On Fire Island everyone uses private disposal systems, and the homes are very close together. The result is that the shallowest water is severely polluted, and most homes are served by a community well driven deep enough to avoid the contamination. Don't expect to get good drinking water from private wells in the densely settled areas on Fire Island.

If you purchase a lot without knowledge of local ordinances and permit requirements, you may later find that your construction costs are increased in order to meet local regulations, or worse, you might require a variance. If you cannot get the variance, it would then be illegal to build on your property. You can avoid much of the red tape that permits generate by purchasing an existing structure. If the property you are considering is far out of line with existing regulations you are probably increasing your risk factor in some area.

6. Building or buying a house near the beach

In reading this book you may conclude that the authors seem to be at cross-purposes. On the one hand, we point out that building on the coast is risky. On the other hand, we provide you with a guide to evaluate the risks, and in this chapter we describe how best to buy or build a house near the beach.

This apparent contradiction is more rational than it might seem at first. For those who will heed the warning, we describe the risks of owning shore-front property. But we realize that coastal development will continue. Some individuals will always be willing to gamble with their fortunes to be near the shore. For those who elect to play this game of real estate roulette, we provide some advice on improving the odds, on reducing (not eliminating) the risks. We do not recommend, however, that you play the game!

If you want to learn more about construction near the beach, we recommend the book *Coastal Design: A Guide for Builders, Planners, and Home Owners* (Van Nostrand Reinhold Company, 1983), which gives more detail on coastal construction and supplements this New York volume. In addition, the Federal Emergency Management Agency's *Design and Construction Manual for Residential Buildings in Coastal High Hazard Areas* is an excellent guide to coastal construction and additional reference material. (See references 87 and 88, appendix C.)

Coastal realty versus coastal reality

Coastal property is not the same as inland property. Do not approach it as if you were buying a lot in a developed woodland or a subdivided farm field. The previous chapters illustrate that the south shore of Long Island, especially the barrier islands, is composed of variable environments and is subjected to nature's most powerful and persistent forces. The reality of the coast is its dynamic character. Property lines are an artificial grid superimposed on this dynamism. If you choose to place yourself or others in this zone, prudence is in order.

A quick glance at the architecture of the structures on the south shore provides convincing evidence that the reality of coastal processes was rarely considered in their construction. Apparently the sea view and aesthetics were the primary considerations. Except for meeting minimal building code requirements, no further thought seems to have been given to the safety of many of these buildings. The failure to follow a few basic architectural guidelines that recognize this reality will have disastrous results in the next major storm.

Life's important decisions are based on an evaluation of the facts. Few of us buy goods, choose a career, or take legal, financial, or medical actions without first evaluating the facts and seeking advice. In the case of coastal property, two general aspects should be evaluated: site safety, and the integrity of the structure relative to the forces to which it will be subjected.

A guide to evaluating the site(s) of your interest is presented below. Following this, the remainder of this chapter focuses on the structure itself, whether cottage or condominium.

The site: checklist for safety evaluation

1. Site elevation is above anticipated storm-surge level.
2. Site is behind a natural protective barrier such as a ridge of sand dunes.
3. Site is in an area of shoreline growth (accretion) or low shoreline erosion. Evidence of an eroding shoreline includes (a) sand bluff or dune scarp—small cliff—at back of beach, (b) mud exposed on beach, (c) threatened man-made structures, and (d) protective devices such as seawalls, groins, or artificially emplaced sand.
4. Site is located on a portion of the island backed by a barrier flat or salt marsh.
5. Site has good access to an evacuation route, which is adequate to handle mass evacuation under storm conditions.
6. Site is away from low, narrow portions of the island.
7. Site is not in an area of historic overwash.
8. Site is well away from migrating inlet.
9. Site is in a vegetated area that suggests stability.
10. Site drains water readily, even in the wet season.

11. Water supply is adequate and uncontaminated. Water and sewage systems are adequate for present demands and anticipated growth.
12. Soil and elevation are suitable for efficient septic tank operation.
13. No compactable layers such as mud are present in soil below footings. (Site is not on a buried salt marsh.)
14. Adjacent buildings are adequately spaced and of sound construction.
15. Federal flood insurance is available.
16. Building codes exist and are really enforced.
17. The year-round residents (who are the electorate) agree with your outlook on the future of the community.

The structure: concept of balanced risk

A certain chance of failure for any structure exists within the constraints of economy and environment. The objective of building design is to create a structure that is both economically feasible and functionally reliable. A house must be affordable and have a reasonable life expectancy free of being damaged, destroyed, or wearing out. In order to obtain such a house, a balance must be achieved among financial, structural, environmental, and other special conditions. Most of these conditions are heightened on the coast—property values are higher, there is a greater desire for aesthetics, the environment is more sensitive, the likelihood of storms is increased, etc.

The individual who builds or buys a home in an exposed area should fully comprehend the risks involved and the chance of harm to home or family. The risks should then be weighed against the benefits to be derived from the residence. Similarly, the developer who is putting up a motel should weigh the possibility of destruction and death during a hurricane versus the money and other advantages to be gained from such a building. Then and only then should construction proceed. For both the homeowner and the developer, proper construction and location reduce the risks involved.

The concept of balanced risk should take into account the following fundamental considerations:

1. Construction must be economically feasible.
2. Because construction must be economically feasible, ultimate and total safety is not obtainable for most homeowners on the coast.
3. A coastal structure, exposed to high winds, waves, or flooding, should be stronger than a structure built inland.
4. A building with a planned long life, such as a year-round residence, should be stronger than a building with a planned short life, such as a mobile home.
5. A building with a high occupancy, such as an apartment building, should be safer than a building with low occupancy, such as a single-family dwelling.
6. A building that houses elderly or sick people should be safer than a building housing able-bodied people.

Structures can be designed and built to resist all but the largest storms and still be within reasonable economic limits.

Structural engineering is the designing and construction of buildings to withstand the forces of nature. It is based on a knowledge of the forces to which the structures will be subjected and an understanding of the strength of building materials. The effectiveness of structural engineering design was reflected in the aftermath of Cyclone Tracy which struck Darwin, Australia, in 1974: 70 percent of housing that was not based on good structural engineering principles was destroyed and 20 percent was seriously damaged—only 10 percent of the housing weathered the storm. In contrast, over 70 percent of the engineered, large commercial, government, and industrial buildings came through with little or no damage, and less than 5 percent of such structures were destroyed. Because housing accounts for more than half of the capital cost of the buildings in Queensland, state government established a building code that requires standardized structural engineering for houses in hurricane-prone areas. This improvement has been achieved with little increase in construction and design costs.

Coastal forces: design requirements

Although northeasters can also be devastating, hurricanes produce the most destructive forces to be reckoned with on the coast (fig. 6.1).

WIND

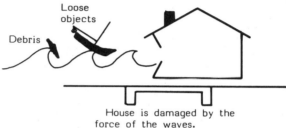

Wind
direction
⇨

Pressure

Pressure

Suction

Suction

Arrows show direction of
forces on house.

DROP IN BAROMETRIC PRESSURE

Low
pressure
outside

Normal high
pressure
inside house

The passing eye of the storm
creates different pressure inside
and out, and high pressure inside
attempts to burst house open.

WAVES

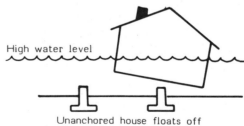

Loose
objects

Debris

House is damaged by the
force of the waves.

HIGH WATER

High water level

Unanchored house floats off
its foundation.

Fig. 6.1. Forces to be reckoned with at the seashore.

Hurricane winds

Hurricane winds can be evaluated in terms of the pressure they exert. A 100-mph wind exerts a pressure or force of about 40 pounds per square foot on a flat surface. The pressure varies with the square of the velocity. For example, a wind of 190-mph velocity exerts a force of 144 pounds per square foot. This force is modified by several factors which must be considered in designing a building. For instance, the effect on a round surface, such as that of a sphere or cylinder, is less than the effect on a flat surface. Also, winds increase with height above ground, so a tall structure is subject to greater pressure than a low structure.

A house or building designed for inland areas is built primarily to resist vertical loads. It is assumed that the foundation and framing must support the load of the walls, floor, and roof, and relatively insignificant horizontal wind forces.

A well-built house in a hurricane-prone area, however, must be constructed to withstand a variety of strong wind forces that may come from any direction. Although many people think that wind damage is caused by uniform horizontal pressures (lateral loads), most damage, in fact, is caused by uplift (vertical), suctional (pressure-differential), and torsional (twisting) forces. High horizontal pressure on the windward side is accompanied by suction on the leeward side. The roof is subject to downward pressure and, more importantly, to uplift. Often a roof is sucked up by the uplift drag of the wind. Usually the failure of houses is in the devices that tie the parts of the structure together. All structural members (beams, rafters, columns) should be fastened together on the assumption that about 25 percent of the vertical load on the member may be a force coming from any direction (sideways or upwards). Such structural integrity is also important if it is likely that the building may be moved to avoid destruction by shoreline retreat.

Storm surge

Storm surge is a rise in sea level above the normal water level during a storm. During hurricanes the coastal zone is inundated by storm surge and accompanying storm waves, and these cause most property damage and loss of life. (For more data on storm surge, see chapter 1.)

Often the pressure of the wind backs water into streams or estuaries already swollen from the exceptional rainfall brought on by the hurricane. Water is piled into the bays between islands and the mainland by the offshore storm. In some cases the direction of flooding may be from the bay side of the island. This flooding is particularly dangerous when the wind pressure keeps the tide from running out of inlets, so that the next normal high tide pushes the accumulated waters back and higher still.

People who have cleaned the mud and contents out of a house subjected to flooding retain vivid memories of its effects. Flooding can cause an unanchored house to float off its foundation and come to rest against another house, severely damaging both. Even if the house itself is left structurally intact, flooding may destroy its contents.

Proper coastal development takes into account the expected level and frequency of storm surge for the area. In general, building standards require that the lowest floor of a dwelling be above the 100-year flood level, with an additional allowance for wave height. At this level, a building has a one-percent probability of being flooded in any given year.

Hurricane waves

Hurricane waves can cause severe damage not only in forcing water onshore to flood buildings but also in throwing boats, barges, piers, houses, and other floating debris inland against standing structures. In addition, waves can destroy coastal structures by scouring away the underlying sand, causing them to collapse. It is possible to design buildings to survive crashing storm surf. Many lighthouses, for example, have survived this. But in the balanced-risk equation, it usually isn't economically feasible to build ordinary cottages to resist the more powerful effects of such forces. On the other hand, cottages can be made considerably more storm-worthy by following the suggestions in the upcoming sections.

The force of a wave may be understood when one considers that a cubic yard of water weighs over three-fourths of a ton; hence, a breaking wave moving shoreward at a speed of several tens of miles per hour can be one of the most destructive elements of a hurricane.

Barometric pressure change

Changes in barometric pressure may also be a minor contributor to structural failure. If a house is sealed at a normal barometric pressure of 30 inches of mercury, and the external pressure suddenly drops to 26.61 inches of mercury (as it did in Hurricane Camille in Mississippi in 1969), the pressure exerted within the house would be 245 pounds per square foot. An ordinary house would explode if it were leakproof. In tornadoes, where there is a severe pressure differential, many houses do just that. In hurricanes the problem is much less severe. Fortunately, most houses leak, but they must leak fast enough to prevent damage. Given the more destructive forces of hurricane wind and waves, pressure differential may be of minor concern. Venting the underside of the roof at the eaves is a common means of equalizing internal and external pressure.

Figure 6.2 illustrates some of the actions that a homeowner can take to deal with the forces just described.

House selection

Some types of houses are better than others at the shore, and an awareness of the differences will help you make a better selection, whether you are building a new house or buying an existing one.

Worst of all are unreinforced masonry houses—whether they be brick, concrete block, hollow clay-tile, or brick veneer—because

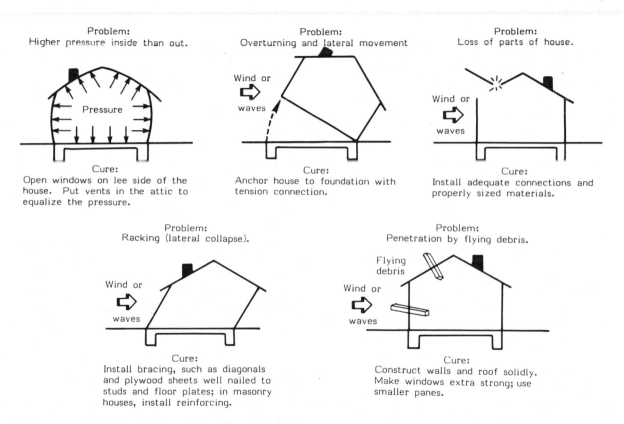

Fig. 6.2. Modes of failure and how to deal with them. Modified from U.S. Civil Defense Preparedness Agency Publication TR-83.

they cannot withstand the lateral forces of wind and wave and the settling of the foundation.

Adequate and extraordinary reinforcing in coastal regions will alleviate the inherent weakness of unit masonry, if done properly. Reinforced concrete and steel frames are excellent but are rarely used in the construction of small residential structures.

It is hard to beat a wood-frame house that is properly braced and anchored and has well-connected members. The well-built wood house will often hold together as a unit, even if moved off its foundations, when other types disintegrate. Although all of the structural types noted above are found in the coastal zone, newer structures tend to be of the elevated wood-frame type.

Keeping dry: pole or "stilt" houses

In coastal areas subject to flooding, nearly all communities have adopted building codes or zoning ordinances that comply with minimum standards established by the National Flood Insurance Program. In V-Zones (see chapter 5), these ordinances or codes generally require that residences be elevated on piling and columns so that the lowest structural member of the lowest floor is at or above the 100-year flood level. Areas below that elevation must be left free of obstruction and must contain no habitable space. If these lowest areas are enclosed by walls, the walls must be designed to break away if struck by waves. The 100-year flood elevation is being adjusted upward in these communities to include wave heights that are superimposed on the storm surge. In A-

Zones, which are areas less vulnerable to direct wave attack, residences can be elevated by any means so that the lowest floor is at or above the elevation of a 100-year flood. Although elevation of a residence by building a mound out of fill generally would be permitted in A-Zones, it is not advisable in many coastal areas where the fill could be eroded by waves or flood currents. In response to these considerations, most modern coastal flood-prone structures are elevated on piling well anchored in the subsoil.

Current building design criteria for pole-house construction under the National Flood Insurance Program is outlined in the *Design and Construction Manual for Residential Buildings in Coastal High Hazard Areas* (see reference 88 in appendix B). Regardless of these requirements, pole type construction with deep embedment of piling is advisable in any area where waves and storm surge will erode foundation material.

Piles. Piles are long, slender columns of wood, steel, or concrete driven into the earth to a sufficient depth to support the vertical load of the house and to withstand the horizontal forces of flowing water, wind, and water-borne debris. Pile construction is especially suitable in areas where scouring—soil "washing out" from under the foundation of a house—is a problem.

Posts. Posts are usually of wood; if of steel, they are called columns. Unlike piles, they are not driven into the ground but, rather, are placed in a predug hole at the bottom of which may be a concrete pad (fig. 6.3). Posts may be held in place by backfilling and tamping earth or by pouring concrete into the hole after the post is in place. Posts are more readily aligned than driven piles

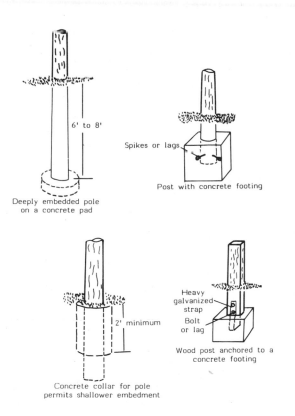

6' to 8'

Deeply embedded pole
on a concrete pad

Spikes or lags

Post with concrete footing

2' minimum

Concrete collar for pole
permits shallower embedment

Heavy
galvanized
strap

Bolt
or lag

Wood post anchored to a
concrete footing

Fig. 6.3. Shallow and deep supports for poles and posts. Source: Southern Pine Association.

and are, therefore, better to use if poles must extend to the roof. In general, treated wood is the cheapest and most common material for both posts and piles.

Piers. Piers are vertical supports, thicker than piles or posts, usually made of reinforced concrete or reinforced masonry (concrete blocks or bricks). They are set on footings and extend to the underside of the floor frame.

Pole construction can be of two types. The poles can be cut off at the first-floor level to support the platform that serves as the dwelling floor. In this case, piles, posts, or piers can be used. Or they can be extended to the roof and rigidly tied into both the floor and the roof. In this way, they become major framing members for the structure and provide better anchorage to the house as a whole (figs. 6.4 and 6.5). A combination of full- and floor-height poles is used in some cases, with the shorter poles restricted to supporting the floor inside the house (fig. 6.6).

Where the foundation material can be eroded by waves or winds, the poles should be deeply embedded and solidly anchored either by driving piles, jetting them, or by drilling deep holes for posts and putting in a concrete pad at the bottom of each hole. Where the embedment is shallow, a concrete collar around the poles improves anchorage (fig. 6.3). The choice depends on the soil conditions. Piles are more difficult than posts to align to match the house frame; posts can be positioned in the holes before backfilling. Inadequate piling depths, improper piling-to-floor connections, and inadequate pile bracing all contribute to structural failure when storm waves liquefy and erode sand support.

Fig. 6.4. Extending poles to the roof, as shown in this photograph, instead of the usual method of cutting them off at the first floor, greatly strengthens a beach cottage. Photograph by Orrin H. Pilkey, Jr.

When post holes are dug, rather than pilings driven, the posts should extend 6 to 10 feet into the ground to provide anchorage. The lower end of the post should rest on a concrete pad, spreading the load to the soil over a greater area to prevent settlement. Where the soil is sandy or the embedment less than, say, 6 feet, it is best to tie the post down to the footing with straps or other anchoring devices. Driven or jetted piles should extend to a depth of 10 feet or more. The NFIP recommends that piling extend to depths of 5 to 10 feet *below* mean sea level.

The floor and the roof should be securely connected to the poles with bolts or other fasteners. When the floor rests on poles that do not extend to the roof, attachment is even more critical. A system of metal straps is often used. Another method is to attach beams to piles with at least two bolts of adequate size. Unfortunately, builders sometimes simply attach the floor beams to a notched pole by one or two undersized bolts. Hurricanes have proven this method insufficient. During the next hurricane on the south shore, many houses will be destroyed because of inadequate attachment.

Local building codes may specify the size, quality, and spacing of the piles, ties, and bracing, as well as the methods of fastening the structure to them. Building codes often are minimal requirements, however, and building inspectors are usually amenable to allowing designs that are equally or more effective.

The space under an elevated house, whether pole-type or otherwise, must be kept free of obstructions in order to minimize the impact of waves and floating debris. If the space is enclosed, the enclosing walls should be designed so that they can break away or fall under flood loads but also remain attached to the house or be heavy enough to sink; thus, the walls cannot float away and add to the water-borne debris problem. Alternative ways of avoiding this problem are designing walls that can be swung up out of the path of the floodwaters, or building them with louvers that allow the water to pass through. The louvered wall is subject to damage from floating debris. The convenience of closing in the ground floor for a garage or extra bedroom may be costly because it may violate local ordinances and result in increased insurance premiums, and actually contribute to the loss of the house in a hurricane. Open wood-lattice work is an acceptable enclosure.

An existing house: what to look for, where to improve

If instead of building a new house, you are selecting a house already built in an area subject to flooding and high winds, consider the following factors: (1) where the house is located; (2) how well the house is built, and (3) how the house can be improved.

Geographic location

Evaluate the site of an existing house using the same principles given earlier for the evaluation of a possible site to build a new house. House elevation, frequency of high water, escape route, and how well the lot drains should be emphasized, but you should go through the complete site-safety checklist given earlier.

You can modify the house after you have purchased it but you can't prevent hurricanes or northeasters. The first step is to stop and consider: Do the pleasures and benefits of this location balance the risks and disadvantages? If not, look elsewhere for a home; if so, then evaluate the house itself.

How well built is the house?

In general, the principles used to evaluate an existing house are the same as those used in building a new one (see references 87 to 99, appendix C).

Before you thoroughly inspect the house in which you are interested, look closely at the adjacent homes. If poorly built, they may float over against your house and damage it in a flood. You may even want to consider the type of people you will have as neighbors: Will they "clear the decks" in preparation for a storm or will they leave items in the yard to become wind-borne missiles?

The house should be well anchored to the ground. If it is simply resting on blocks, rising water may cause it to float off its foundation and come to rest against your neighbor's house or out in the middle of the street. If well built and well braced internally, it may be possible to move the house back to its proper location, but chances are great that the house will be too damaged to be habitable.

If the house is on piles, posts, or poles, check to see if the floor beams are adequately bolted to them. If it rests on piers, crawl under the house if space permits to see if the floor beams are securely connected to the foundation. If the floor system rests unanchored on piers, do not buy the house.

It is difficult to discern whether a house built on a concrete slab is properly bolted to the slab because the inside and outside walls hide the bolts. If you can locate the builder, ask if such bolting

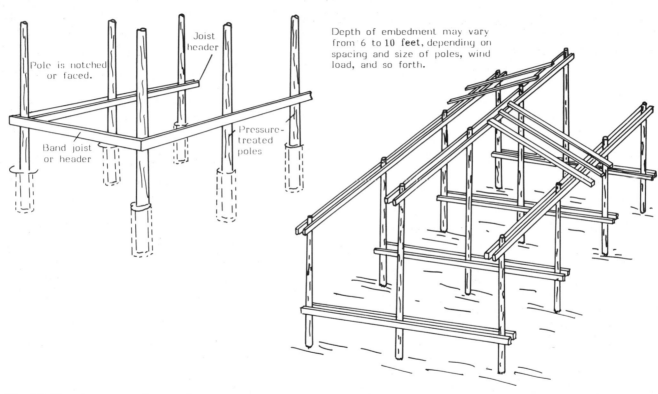

Joist header

Pole is notched or faced.

Band joist or header

Pressure-treated poles

Depth of embedment may vary from 6 to 10 feet, depending on spacing and size of poles, wind load, and so forth.

Fig. 6.5. Framing system for an elevated house. Source: Southern Pine Association.

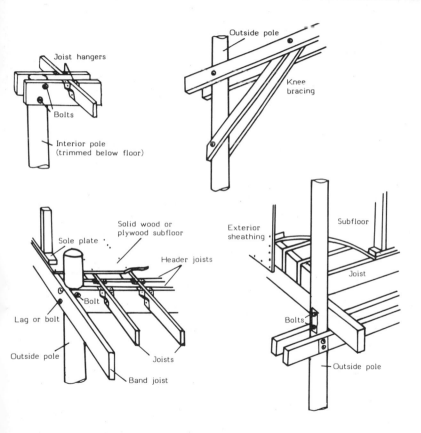

Fig. 6.6. Tying floors to poles. Source: Southern Pine Association.

was done. Better yet, if you can get assurance that construction of the house complied with the provisions of a building code serving the needs of that particular region, you can be reasonably sure that all parts of the house are well anchored: the foundation to the ground, the floor to the foundation, the walls to the floor, and the roof to the walls (figs. 6.7, 6.8, and 6.9).

Be aware that many builders, carpenters, and building inspectors who are accustomed to traditional construction are apt to regard metal connectors, collar beams, and other such devices as newfangled and unnecessary. If consulted, they may assure you that a house is as solid as a rock, when in fact, it is far from it. Nevertheless, it is wise to consult the builder or knowledgeable neighbors when possible.

The roof should be well anchored to the walls. This will prevent uplifting and separation from the walls. Visit the attic to see if such anchoring exists. Simple toe-nailing (nailing at an angle) is not adequate; metal fasteners are needed. Depending on the type of construction and the amount of insulation laid on the floor of the attic, these may or may not be easy to see. If roof trusses or braced rafters were used, it should be easy to see whether the various members, such as the diagonals, are well fastened together. Again, simple toe-nailing will not suffice. Some builders, unfortunately, nail parts of a roof truss just enough to hold it together to get it in place. A collar beam or gusset at the peak of the roof (fig. 6.10) provides some assurance of good construction.

As for framing, the fundamental rule to remember is that all structural elements should be fastened together and anchored to

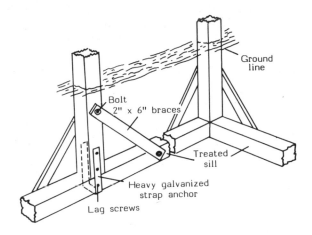

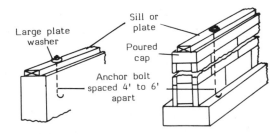

Fig. 6.7. Foundation anchorage. Top: anchored sill for shallow embedment. Bottom: anchoring sill or plate to foundation. Source of bottom drawing: *Houses Can Resist Hurricanes*, U.S. Forest Service Research Paper FPL 33.

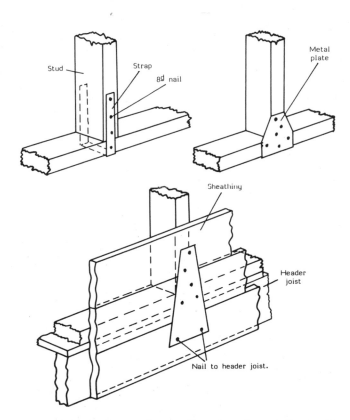

Fig. 6.8. Stud-to-floor, plate-to-floor framing methods. Source: *Houses Can Resist Hurricanes*, U.S. Forest Service Research Paper FPL 33.

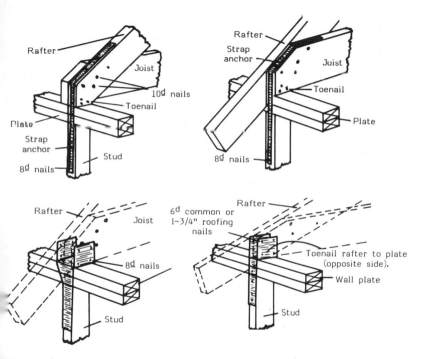

Fig. 6.9. Roof-to-wall connections. The top drawings show metal strap connectors: left, rafter to stud; right, joist to stud. The bottom left drawing shows a double-member metal-plate connector—in this case with the joist to the right of the rafter. The bottom right drawing shows a single-member metal-plate connector. Source: *Houses Can Resist Hurricanes*, U.S. Forest Service Research Paper FPL 33.

the ground In such a manner as to resist all forces, regardless of which direction these forces may come from. This prevents overturning, floating off, racking, or disintegration.

Quality roofing material should be well anchored to the sheathing. A poor roof covering will be destroyed by hurricane-force winds, allowing rain to enter the house and damage ceilings, walls, and the contents of the house. Galvanized nails (two per shingle) should be used to connect wood shingles and shakes to wood sheathing, and should be long enough to penetrate through the sheathing (fig. 6.10). Threaded nails should be used for plywood sheathing. For roof slopes that rise 1 foot for every 3 feet or more of horizontal distance, exposure of the shingle should be about one-fourth of its length (4 inches for a 16-inch shingle). If shakes (thicker and longer than shingles) are used, less than one-third of their length should be exposed.

In hurricane areas, asphalt shingles should be exposed somewhat less than usual. A mastic or seal-tab type or an interlocking shingle of heavy grade should be used. A roof underlay of asphalt-saturated felt and galvanized roofing nails or approved staples (six for each three-tab strip) should be used.

The shape of the house is important. A hip roof, which slopes in four directions, is better able to resist high winds than a gable roof, which slopes in two directions. This was found to be true in Hurricane Camille in 1969 in Mississippi and, later, in Cyclone Tracy, which devastated Darwin, Australia, in December 1974. The reason is twofold: the hip roof offers a smaller shape for the

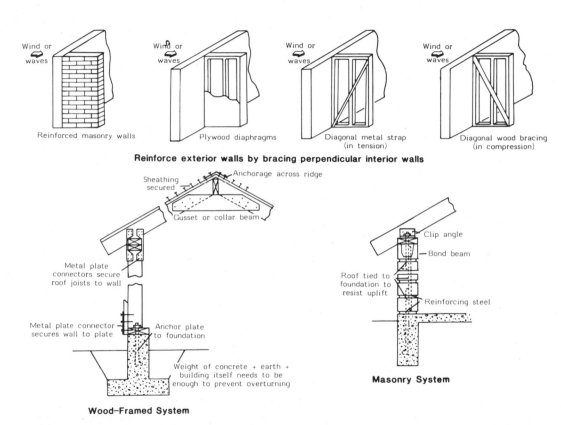

Fig. 6.10. Where to strengthen a house. Modified from U.S. Civil Defense Preparedness Agency Publication TR-83.

wind to blow against, and its structure is such that it is better braced in all directions.

Note also the horizontal cross section of the house (the shape of the house as viewed from above). The pressure exerted by a wind on a round or elliptical shape is about 60 percent of that exerted on the common square or rectangular shape; the pressure exerted on a hexagonal or octagonal cross section is about 80 percent of that exerted on a square or rectangular cross section.

The design of a house or building in a coastal area should minimize structural discontinuities and irregularities. A house should have a minimum of nooks and crannies and offsets on the exterior, because damage to a structure tends to concentrate at these points. Some of the newer beach cottages along the American coast are of a highly angular design with such nooks and crannies. Award-winning architecture will be a storm loser if the design has not incorporated the technology for maximizing structural integrity with respect to storm forces. When irregularities are absent, the house reacts to storm winds as a complete unit.

Brick, concrete-block, and masonry-wall houses should be adequately reinforced. This reinforcement is hidden from view. Building codes applicable to high-wind areas often specify the type of mortar, reinforcing, and anchoring to be used in construction. If you can get assurance that the house was built in compliance with a building code designed for such an area, consider buying it. At all costs, avoid unreinforced masonry houses. Even if reinforced, masonry structures are not recommended in V-Zones.

A poured-concrete bond beam at the top of the wall just under the roof is one indication that the house is well built (fig. 6.11). Most bond beams are formed by putting in reinforcing and pouring concrete in U-shaped concrete blocks. From the outside, however, you can't distinguish these U-shaped blocks from ordinary ones and therefore can't be certain that a bond beam exists; the vertical reinforcing should penetrate the bond beam.

Some architects and builders use a stacked bond (one block directly above another), rather than overlapped or staggered blocks, because they believe it looks better. The stacked bond is definitely weaker than the latter. Unless you have proof that the walls are adequately reinforced to overcome this lack of strength, you should avoid this type of construction.

In past hurricanes, the brick veneer of many houses has separated from the wood frame, even when the houses remained standing. Asbestos-type outer-wall panels used on many houses in Darwin, Australia, were found to be brittle and they broke up under the impact of wind-borne debris in Cyclone Tracy. Both types of construction should be avoided along the coast.

Windows and large glass areas, especially those that face the ocean, should be protected. Wind-blown sand can very quickly frost a window and thereby decrease its aesthetic value. More seriously, an object blown through a window during a storm causes dangerous flying glass. Both of these problems can be avoided if the house has shutters. Check to see if it does, and if they are functional.

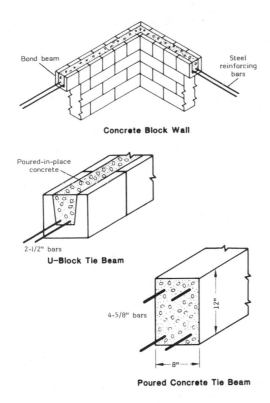

Bond beam

Steel reinforcing bars

Concrete Block Wall

Poured-in-place concrete

2-1/2" bars

U-Block Tie Beam

4-5/8" bars

12"

8"

Poured Concrete Tie Beam

Fig. 6.11. Reinforced tie beam (bond beam) for concrete block walls —to be used at each floor level and at roof level around the perimeter of the exterior walls.

Consult a good architect or structural engineer for advice if you are in doubt about any aspects of a house. A few dollars spent for wise counsel may save you from later financial grief.

To summarize, a beach house should have (1) roof tied to walls, walls tied to foundation, and foundation anchored to the earth (the connections are potentially the weakest link in the structural system); (2) a shape that resists storm forces; (3) shutters for all windows, especially those facing the ocean; (4) floors high enough (sufficient elevation) to be above most storm waters (usually the 100-year stillwater flood level plus 2 to 6 feet to account for wave height); (5) piles that are of sufficient depth or embedded in concrete to anchor the structure and to withstand erosion; and (6) piling that is well braced.

What can be done to improve an existing house?

If you presently own a house or are contemplating buying one in a hurricane-prone area, you will want to know how to improve occupant protection in the house. If so, you should obtain the excellent publication, *Wind-Resistant Design Concepts for Residences*, by Delbart B. Ward (reference 90, appendix C). Of particular interest are the sections on building a refuge shelter module within a residence. Also noteworthy are two supplements to this publication (reference 91, appendix C) which deal with buildings larger than single-family residences and which may be of interest to the general public, especially residents in urban areas. These provide a means of checking whether the responsible authorities are doing their jobs to protect schools, office buildings, and apartments. Several other pertinent references are listed in appendix C.

Suppose your house is resting on blocks but not fastened to them and, thus, is not adequately anchored to the ground. Can anything be done? One solution is to treat the house like a mobile home by screwing ground anchors into the ground to a depth of 4 feet or more and fastening them to the underside of the floor systems.

Calculations to determine the needed number of ground anchors will differ between a house and a mobile home, because each is affected differently by the forces of wind and water. Recent practice is to put the commercial steel-rod anchors in at an angle in order to align them better with the direction of the pull. If a vertical anchor is used, the top 18 inches or so should be encased in a concrete cylinder about 12 inches in diameter. This prevents the top of the anchor rod from bending or slicing through the wet soil from the horizontal component of the pull.

Diagonal struts, either timber or pipe, may also be used to anchor a house that rests on blocks. This is done by fastening the upper ends of the struts to the floor system, and the lower ends to individual concrete footings substantially below the surface of the ground. These struts must be able to take both uplift (tension) and compression, and should be tied into the concrete footing with anchoring devices such as straps or spikes.

If the house has a porch with exposed columns or posts, it should be possible to install tiedown anchors on their tops and bottoms. Steel straps should suffice in most cases.

When accessible, roof rafters and trusses should be anchored to the wall system. Usually the roof trusses or braced rafters are sufficiently exposed to make it possible to strengthen joints (where two or more members meet) with collar beams or gussets, particularly at the peak of the roof (fig. 6.10).

A competent carpenter, architect, or structural engineer can review the house with you and help you decide what modifications are most practical and effective. Do not be misled by someone who is resistant to new ideas. One builder told a homeowner, "You don't want all those newfangled straps and anchoring devices. If you use them, the whole house will blow away, but if you build in the usual manner [with members lightly connected], you may lose only part of it."

In fact, the very purpose of the straps is to prevent any or all of the house from blowing away. The Standard Building Code (previously known as the Southern Standard Building Code and still frequently referred to by that name) says, "Lateral support securely anchored to all walls provides the best and only sound structural stability against horizontal thrusts, such as winds of exceptional velocity" (see reference 92, appendix C). And the cost of connecting all elements securely adds very little to the cost of the frame of the dwelling, usually under 10 percent, and a very much smaller percentage to the total cost of the house.

If the house has an overhanging eave and there are no openings on its underside, it may be feasible to cut openings and screen them. These openings keep the attic cooler (a plus in the summer) and help to equalize the pressure inside and outside of the house during a storm with a low-pressure center.

Another way a house can be improved is to modify one room so that it can be used as an emergency refuge in case you are trapped in a major storm. (This is *not* an alternative to evacuation prior to a hurricane.) Examine the house and select the best room to stay in during a storm. A small windowless room such as a bathroom, utility room, den, or storage space is usually stronger than a room with windows. A sturdy inner room, with more than one wall between it and the outside, is safest. The fewer doors, the better; an adjoining wall or baffle-wall shielding the door adds to the protection.

Consider bracing or strengthening the interior walls. Such reinforcement may require removing the surface covering and installing plywood sheathing or strap bracing. Where wall studs are exposed, bracing straps offer a simple way to achieve needed reinforcement against the wind. These straps are commercially produced and are made of 16-gauge galvanized metal with prepunched holes for nailing. These should be secured to studs and wall plates as nail holes permit (fig. 6.10). Bear in mind that they are good only for tension.

If, after reading this, you agree that something should be done to your house, do it now. Do not put it off until the next hurricane or northeaster hits you!

High-rise buildings: the urban shore

A high-rise building on the beach is generally designed by an architect and a structural engineer who are presumably well qualified and aware of the requirements for building on the shoreline. Tenants of such a building, however, should not assume that it is therefore invulnerable. Many people living in apartment buildings of two or three stories were killed when the buildings were destroyed by Hurricane Camille in Mississippi in 1969. Storms have destroyed the fronts of five-story buildings in Delaware. Larger high rises have yet to be thoroughly tested by a major hurricane.

The first aspect of high-rise construction that a prospective apartment dweller or condo owner must consider is the type of piling used. High rises near the beach should be built so that even if the foundation is severely undercut during a storm the building will remain standing. It is well known in construction circles that shortcuts are sometimes taken by the less scrupulous builders, and piling is not driven deeply enough. Just as important as driving the piling deep enough to resist scouring and to support the loads they must carry is the need to fasten piles securely to the structure above them which they support. The connections must resist horizontal loads from wind and wave during a storm and also uplift from the same sources. It is a joint responsibility of builders and building inspectors to make sure the job is done right. In 1975 Hurricane Eloise exposed the foundation of a just-under-construction high rise in Florida, revealing that some of the piling was not attached to the building. This happened in Panama City, Florida, but such problems probably exist everywhere that high rises crowd the beach.

Despite the assurances that come with an engineered structure, life in a high-rise building holds definite drawbacks that prospective tenants should take into consideration. The negative conditions that must be evaluated stem from high wind, high water, and poor foundations.

Pressure from the wind is greater near the shore than it is inland, and it increases with height. If you are living inland in a two-story house and plan to move to the eleventh floor of a high rise on the shore, you should expect five times more wind pressure than you are accustomed to. This can be a great and possibly devastating surprise.

The high wind-pressure actually can cause unpleasant motion of the building. It is worthwhile to check with current residents of a high rise to find out if it has undesirable motion characteristics; some have claimed that the swaying is great enough to cause motion sickness. More seriously, high winds can break windows and damage other property, and of course they can hurt people. Tenants of severely damaged buildings will have to relocate until repairs are made.

Those who are interested in researching the subject further— even the knowledgeable engineer or architect who is engaged to design a structure near the shore—should obtain a copy of *Structural Failures: Modes, Causes, Responsibilities* (reference 93, appendix C). Of particular importance is the chapter entitled, "Failure of Structures Due to Extreme Winds." This chapter analyzes wind damage to engineered high-rise buildings from the storms at Lubbock and Corpus Christi, Texas, in 1970.

Another occurrence that affects a multifamily, high-rise building more seriously than a low-occupancy structure is a power failure or blackout. Such an occurrence is more likely along the coast than inland because of the more severe weather conditions associated with coastal storms. A power failure can cause great distress. People can be caught between floors in an elevator. New York City had that experience not too long ago. Think of the mental and physical distress after several hours of confinement. And compound this with the roaring winds of a hurricane whipping around the building, sounding like a freight train. In this age of electricity, it is easy to imagine many of the inconveniences that can be caused by a power failure in a multistory building.

Fire is an extra hazard in a high-rise building. Even recently constructed buildings seem to have difficulties. The television pictures of a woman leaping from the window of a burning building in New Orleans rather than be incinerated in the blaze are a horrible reminder from recent history. The number of hotel fires of the last few years demonstrates the problems. Fire department equipment reaches only so high. And many areas along the coast are too sparsely populated to afford high-reaching equipment.

Fire and smoke travel along ventilation ducts, elevator shafts, corridors, and similar passages. The situation *can be* corrected and the building made safer, especially if it is new. Sprinkler systems should be operated by gravity water systems rather than by powered pumps (because of possible power failure). Gravity systems use water from tanks higher up in the building. Battery-operated emergency lights that come on only when the other lights fail, better fire walls and automatic sealing doors, pressurized stairwells, and emergency-operated elevators in pressurized shafts will all contribute to greater safety. Unfortunately all of these improvements cost money, and that is why they are often omitted unless required by the building code.

There are two interesting reports on damage caused by Hurricane Eloise, which struck the Florida Panhandle the morning of September 23, 1975. One is by Herbert S. Saffir, a Florida consulting engineer; the other is by Bryon Spangler of the University of Florida. The forward movement of the hurricane was unusually fast, causing its duration in a specific area to be lessened, thus minimizing damage from both wind and tidal surge. The still-water height at Panama City was 16 feet above mean sea level, with 3-foot waves and wind gusts of 154 mph.

At least one-third of the older structures in the Panama City area collapsed. These were beachfront motels, restaurants, apartments and condominium complexes, and some permanent residences. The structures built on piling survived with minimal damage. In one case, part of a motel was on spread footings and part on piles. Just the part on spread footings was severely damaged.

The anchorage systems, connection between concrete piles or concrete piers and the grade beams under several high-rise buildings were inadequate to resist uplift loads, illustrating that code enforcement or proper inspection by the design engineer is essential.

Many of the residences and some of the buildings were built on spread footings which failed because the sand they were resting on washed away with scour. Failure of the footings resulted in failure of the superstructure.

Some of the high-rise buildings suffered glass damage in both windows and sliding glass doors.

Apparently few of the residences and buildings were built to conform to the Standard Building Code, as many of them existed before the code's implementation. If such requirements had been met, much of the damage could have been prevented.

Some surprising things were noticed. In almost every case where there was a swimming pool, considerable erosion occurred. Loss of sand beneath the walkways prior to the storm created a channel for the water to flow through and wash out more sand during the storm, which in turn increased both the velocity and quantity of the flow of water in the channel. This ate away the sand supporting adjacent structures, accelerating their failures.

Slabs on grade performed poorly. Often wave action washed out the sand underneath the slab. When this occurred there was no longer support for the structure above it, and failure resulted.

The storm revealed some shoddy construction. Some builders had placed wire mesh for a slab directly on the sand. Then the concrete was poured on top of it, leaving the mesh below and in the sand, where it served no structural purpose. To be effective, it should have been set on blocks or chairs, or pulled up into the slab during the pouring of the concrete.

In some cases cantilevered slabs, for projecting overhangs, were reinforced for the usual downward gravity loads. Unfortunately when waves dashed against the buildings they splashed upward, imposing an upward force against the slab for which it was not reinforced, causing it to crack and fail.

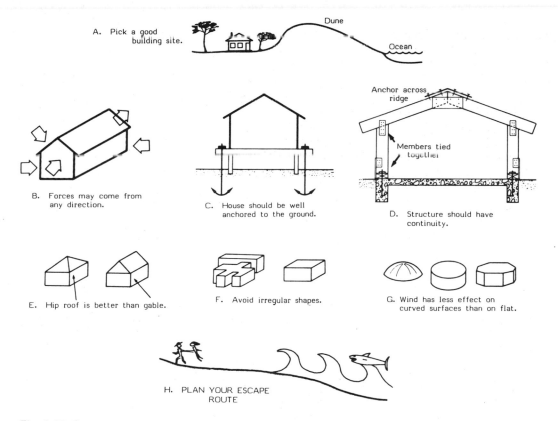

A. Pick a good building site.

B. Forces may come from any direction.

C. House should be well anchored to the ground.

D. Structure should have continuity.

E. Hip roof is better than gable.

F. Avoid irregular shapes.

G. Wind has less effect on curved surfaces than on flat.

H. PLAN YOUR ESCAPE ROUTE

Fig. 6.12. Some rules in selecting or designing a house.

An unending game: only the players change

Hurricane or calm, receding shore or accreting land, storm-surge flood or sunny sky, migrating dune or maritime forest, win or lose, the gamble of coastal development will continue. If you choose your site with natural safety in view, follow good structural engineering design in construction, and take a generally prudent approach to living at the shore (fig. 6.12), then you become the gambler who knows when to hold them, when to fold them, and when to walk away.

Our goal is to provide guidance to today's and tomorrow's players. This book is not the last or by any means the complete guide to coastal living, but it should provide a beginning. In the appendices that follow are additional resources that we hope interested readers will pursue.

Appendix A. Hurricane checklist

Keep this checklist handy for protection of family and property.

When a hurricane threatens

__ Listen for official weather reports.

__ Read your newspaper and listen to radio and television for official announcements.

__ Note the address of the nearest emergency shelter.

__ Know the official evacuation route in advance.

__ Pregnant women, the ill, and the infirm should call a physician for advice.

__ Be prepared to turn off gas, water, and electricity where it enters your home.

__ Fill tubs and containers with water (one quart per person per day).

__ Make sure your car's gas tank is full.

__ Secure your boat. Use long lines to allow for rising water.

__ Secure movable objects on your property:

__ doors	__ hoses
__ outdoor furniture	__ gates
__ shutters	__ garbage cans
__ garden tools	__ other

__ Board up or tape windows and glassed areas. Draw drapes and window blinds across windows and glass doors. Remove furniture in their vicinity.

__ Stock adequate supplies:

__ transistor radio	__ flashlights
__ fresh batteries	__ candles
__ canned heat	__ matches
__ hammer	__ nails
__ boards	__ screwdriver
__ pliers	__ ax*
__ hunting knife	__ rope*
__ tape	__ plastic drop cloths, waterproof bags, ties
__ first-aid kit	
__ prescribed medicines	__ containers for water
__ water purification tablets	__ disinfectant
__ insect repellent	__ canned food, juices, soft drinks (see below)
__ gum, candy	
__ life jackets	__ hard-top head gear
__ charcoal bucket and charcoal	__ fire extinguisher
__ buckets of sand	__ can opener and utensils

*Take an ax (to cut an emergency escape opening) if you go to the upper floors or attic of your home. Take rope for escape to ground when water subsides.

Suggested storm food stock for family of four

__ two 13-oz. cans evaporated milk
__ four 7-oz. cans fruit juice
__ 2 cans tuna, sardines, Spam, chicken
__ three 10-oz. cans vegetable soup
__ 1 small can of cocoa or Ovaltine
__ one 15-oz. box raisins or prunes
__ salt
__ pet food?
__ one 14-oz. can cream of wheat or oatmeal
__ one 8-oz. jar peanut butter or cheese spread
__ two 16-oz. cans pork and beans
__ one 2-oz. jar instant coffee or tea
__ 2 packages of crackers
__ 2 pounds of sugar
__ 2 quarts of water per person

Special precautions for apartments/condominiums

__ Make one person the building captain to supervise storm preparation.
__ Know your exits.
__ Count stairs on exits; you'll be evacuating in darkness.
__ Locate safest areas for occupants to congregate.
__ Close, lock, and tape windows.
__ Remove loose items from terraces (and from your absent neighbors' terraces).

__ Remove or tie down loose objects from balconies or porches.
__ Assume other trapped people may wish to use the building for shelter.

Special precautions for businesses

__ Take photos of building and merchandise.
__ Assemble insurance policies.
__ Move merchandise away from plate glass.
__ Move merchandise to as high a location as possible.
__ Cover merchandise with tarps or plastic.
__ Remove outside display racks and loose signs.
__ Take out lower file drawers, wrap in trash bags, and store high.
__ Sandbag spaces that may leak.
__ Take special precautions with reactive or toxic chemicals.

If you remain at home

__ Stay indoors. Remember, the first calm may be the hurricane's eye. Remain indoors until an official all-clear is given.
__ Stay on the "downwind" side of the house. Change your position as the wind changes.
__ If your house has an "inside" room, it may be the most secure part of the structure.
__ Keep continuous communications watch for *official* information on radio and television.
__ Keep calm. Your ability to meet emergencies will help others.

If evacuation is advised

— Leave as soon as you can. Follow official instructions only.
— Follow official evacuation routes unless those in authority direct you to do otherwise.
— Take these supplies:
    — change of warm, protective clothes
    — first-aid kit
    — baby formula
    — identification tags: include name, address, and next of kin (wear them)
    — flashlight
    — food, water, gum, candy
    — rope, hunting knife
    — waterproof bags and ties
    — can opener and utensils
    — disposable diapers
    — special medicine
    — blankets and pillows in waterproof casings
    — radio
    — fresh batteries
    — bottled water
    — purse, wallet, valuables
    — life jackets
    — games and amusements for children
— Disconnect all electric appliances except refrigerator and freezer. Their controls should be turned to the coldest setting and the doors kept closed.
— Leave food and water for pets. Seeing-eye dogs are the only animals allowed in the shelters.
— Shut off water at the main valve (where it enters your home).
— Lock windows and doors.
— Keep important papers with you:
    — driver's license and other identification
    — insurance policies
    — property inventory
    — Medic Alert or other device to convey special medical information

During the hurricane

— Stay indoors and away from windows and glassed areas.
— If you are advised to evacuate, **do so at once**.
— Listen for continuing weather bulletins and official reports.
— Use your telephone only in an emergency.
— Follow official instructions only. Ignore rumors.
— Keep **open** a window or door on the side of the house opposite the storm winds.
— Beware of the **"eye of the hurricane."** A lull in the winds is not an indication that the storm has passed. Remain indoors unless emergency repairs are necessary. Exercise caution. Winds may resume suddenly, in the opposite direction and with greater

force than before. Remember, if wind direction does change, the open window or door must be changed accordingly.

__ Be alert for rising water.

__ If electric service is interrupted, note the time.

    __ Turn off major appliances, especially air conditioners.

    __ Do not disconnect refrigerators or freezers. Their controls should be turned to the coldest setting and doors closed to preserve food as long as possible.

    __ Keep away from fallen wires. Report location of such wires to the utility company.

__ If you detect **gas**:

    __ Do not light matches or turn on electrical equipment.

    __ Extinguish all flames.

    __ Shut off gas supply at the meter.*

    __ Report gas service interruptions to the gas company.

Water:

    __ The only **safe** water is the water you stored before it had a chance to come in contact with flood waters.

    __ Should you require an additional supply, be sure to boil water for 30 minutes before use.

    __ If you are unable to boil water, treat water you will need with water purification tablets.

Note: An official announcement will proclaim tap water "safe."

*Gas should be turned back on only by a gas serviceman or licensed plumber.

Treat all water except stored water until you hear the announcement.

After the hurricane has passed

__ Listen for official word of danger having passed.

__ Watch out for loose or hanging power lines as well as gas leaks. People have survived storms only to be electrocuted or burned. Fire protection may be nil because of broken power lines.

__ Walk or drive carefully through the storm-damaged area. Streets will be dangerous because of debris, undermining by washout, and weakened bridges. Watch out for poisonous snakes and insects driven out by flood waters.

__ Eat nothing and drink nothing that has been touched by flood waters.

__ Place spoiled food in plastic bags and tie securely.

__ Dispose of all mattresses, pillows, and cushions that have been in flood waters.

__ Contact relatives as soon as possible.

Note: If you are stranded, signal for help by waving a flashlight at night or white cloth during the day.

Appendix B. A guide to government agencies involved in coastal development

Numerous agencies at all levels of government are engaged in planning, regulating, or studying coastal development in New York. These agencies issue permits for various phases of construction and provide information on development to the homeowner, developer, or planner. Following is an alphabetical list of topics related to coastal development; under each topic are the names of agencies to consult for information on that topic.

Aerial photography and maps

Comparison of aerial photographs with maps is one of the most widely used techniques to document shoreline changes. Settlements of legal disputes involving changes in shorelines frequently rely heavily on this type of documentation. Topographic maps are available from:

Distribution Section
U.S. Geological Survey
1200 South Eads Street
Arlington, VA 22202

An index map of New York State (available free) lists the maps available, and gives names of local suppliers and details for ordering maps.

In addition, the New York State Library in Albany has a complete collection of topographic maps as well as historical maps dating from the mid-seventeenth century. For additional information contact:

New York State Library
Manuscripts and Special Collections
Cultural Education Center
Empire State Plaza
Albany, NY 12230
Phone: (518) 474-4461

Aerial photographs are available from government and private sources. In general the private sources, although more expensive, are considerably speedier in supplying you with copies. A list of suppliers of Long Island photographs is given in reference 2, appendix C.

For information on flood-zone maps, see *Insurance*.

Nautical charts in several scales contain navigation information and bottom depths. An index map for these charts is available from:

National Ocean Survey
Distribution Division (C-44)
N.O.A.A.
Riverdale, MD 20840
Phone: (301) 436-6990

Beach erosion

Information on beach erosion, inlet migration, and floods is available from:

> Coastal Engineering Branch
> New York District Army Corps of Engineers
> 26 Federal Plaza
> New York, NY 10007
> Phone: (212) 264-9084

> Coastal Management Section
> New York State Department of Environmental Conservation
> 50 Wolf Road
> Albany, NY 12233
> Phone: (518) 457-4208

Information on Long Beach Island is available from:

> Department of Conservation and Waterways
> Town of Hempstead
> Lido Blvd.
> Point Lookout, NY 11569
> Phone: (516) 431-9200

Building permits

A variety of permits may be required depending upon the construction site you select. The process is further complicated by overlapping jurisdictions and special requirements when building in the coastal zone. Probably the best general advice is to contact the following office for help in guiding you through the process of acquiring permits:

> Chief Permit Administrator
> New York State Department of Environmental Conservation
> Department of Regulatory Affairs
> Building 40, State University of New York at Stony Brook
> Stony Brook, NY 11794
> Phone: (516) 751-7900

Civil disaster and preparedness assistance. See also *Insurance*.

The coordinating office for all emergency services on Long Island is:

> Suffolk County Department of Emergency Preparedness
> P.O. Box 127
> Yaphank Avenue
> Yaphank, NY 11980
> Phone: (516) 924-4400

Immediate temporary aid in case of storm or fire can be obtained from:

> American Red Cross
> 475 Main Street
> Patchogue, NY 11772
> Phone: (516) 475-6202

Federal Emergency Management Agency
Region II
26 Federal Plaza, Room 1349
New York, NY 10278
Phone: (212) 264-8980

American National Red Cross
Disaster Services
Washington, DC 20006
Phone: (202) 857-3722

Coastal zone management program

The coastal zone management program in New York has recently been accepted into the federal program, and this unit will now have a greater impact on coastal dwellers.

For current information on the state of the management program contact:

Robert C. Hansen
Coastal Program Director
Department of State
162 Washington Avenue
Albany, NY 12231

Specific questions about the coastal erosion aspects of the program should be addressed to:

William W. Daley
Chief, Coastal Erosion Division
New York State Department of Environmental Conservation
50 Wolf Road
Albany, NY 12233

Dredging, filling, and construction in coastal waterways

New York State requires that all those who wish to dredge, fill, or otherwise alter marshlands, bay bottoms, or tidelands apply for a permit from the State Department of Environmental Conservation. For information write or call:

New York State Department of Environmental Conservation
Department of Regulatory Affairs
Bldg. 40, State University of New York at Stony Brook
Stony Brook, NY 11794
Phone: (516) 751-7900

In addition, federal law requires a permit for those who wish to dredge, fill, or place any structure in navigable waters (almost any marine waters). Application for a permit or information concerning the permit should be addressed to:

Chief of the Regulatory Branch
Operations Division
New York District Army Corps of Engineers
26 Federal Plaza
New York, NY
Phone: (212) 264-3996

Additional permits are required by some local municipalities, but the state D.E.C. office listed above can give you direction on local regulations.

In Suffolk County additional information is available from:

Waterways Division
Suffolk County Department of Public Works
Yaphank Avenue
Yaphank, NY 11980
Phone: (516) 924-4300

General coastal information

For information on state activities contact:

New York State Sea Grant Institute
411 State Street
Albany, NY 12246

For information on Long Island contact:

Sea Grant Extension Program
Bldg. H, South Campus
State University of New York at Stony Brook
Stony Brook, NY 11794
Phone: (516) 246-7778

Marine Sciences Research Center
State University of New York at Stony Brook
Stony Brook, NY 11794
Phone: (516) 246-6543

Geology. See Water resources.

Hurricane information

The National Oceanic and Atmospheric Administration is the best agency from which to request information on hurricanes. NOAA storm evacuation maps are prepared for vulnerable areas and cost $2.00 each. To find out whether a map is available for your area, call or write:

Distribution Division (C-44)
National Ocean Survey
National Oceanic and Atmospheric Administration
Riverdale, MD 20840
Phone: (301) 436-8194

Insurance

In coastal areas special building requirements must often be met in order to obtain flood or windstorm insurance. To find out the requirements for your area, check with your insurance agent. Further information is available from:

National Flood Insurance Program
Federal Insurance Administration
Federal Emergency Management Agency
Washington, DC 20472

Federal Emergency Management Agency
Region II
26 Federal Plaza, Room 1349
New York, NY 10278
Phone: (212) 264-8980

For V-Zone coverage or request for individual structure rating contact:

National Flood Insurance Program
Attn: V-Zone Underwriting Specialist
6430 Rockledge Drive
Bethesda, MD 20817
Phone: (800) 638-6620 (toll free)

For copies of postconstruction elevation certificate contact:

National Flood Insurance Program
Forms Order Unit
P.O. Box 34294
Bethesda, MD 20817

Your insurance agent or community building inspector should be able to provide you with information about the location of your building site on the Federal Insurance Rate Map (FIRM), and the elevation required for the first floor to be above the 100-year flood level. If they cannot provide this information, request the FIRM for your area from the FEMA Region II office. Note that a flood policy under the national flood insurance program is separate from your regular homeowner's policy.

Parks and recreation

A large portion of the south shore has been set aside for parks and recreation areas. Town and village beaches require that you display a resident parking permit on your car during the peak tourist months of July and August, but state and federal facilities are open to all. For additional information on services available you can contact the following addresses.

For information on federal facilities contact:

Gateway National Recreation Area
Headquarters Bldg. No. 69
Floyd Bennett Field
Brooklyn, NY 11234
Phone: (212) 630-0393

Fire Island National Seashore
120 Laurel Avenue
Patchogue, NY 11772
Phone: (516) 289-4810

For information on state facilities contact:

Long Island State Park and Recreation Commission
Belmont Lake State Park
Box 247
Babylon, NY 11702
Phone: (516) 669-1000

For information on county facilities contact:

Nassau County Department of Recreation and Parks
Public Information
Recreation Administration Bldg.
Eisenhower Park
East Meadow, NY 11554
Phone: (516) 292-4200

Suffolk County Department of Parks
P.O. Box 144
West Sayville, NY 11796
Phone: (516) 567-1700

For information on town facilities contact:

Town of East Hampton Department of Recreation
159 Pantigo Road
East Hampton, NY 11937
Phone: (516) 267-8420

Town of Southampton
Department of Recreation
Jackson Avenue
Hampton Bays, NY 11946
Phone: (516) 728-4554

Town of Islip Department of Parks,
   Recreation and Cultural Affairs
50 Irish Lane
East Islip, NY 11751
Phone: (516) 224-5400

Town of Babylon
Department of Parks and Recreation
151 Phelps Lane
North Babylon, NY 11703
Phone: (516) 957-3107

Town of Oyster Bay
Parks and Recreation
800 South Oyster Bay Road
Hicksville, NY 11801
Phone: (516) 922-5800

Town of Hempstead
Department of Parks and Recreation
Hempstead Executive Plaza
50 Clinton Street
Hempstead, NY 11550
Phone: (516) 489-5000

Planning

There are many planning boards for towns and municipalities on Long Island. Begin by contacting the following:

Principal Planner
Long Island Regional Planning Board
Veterans Memorial Highway
Hauppauge, NY 11787
Phone: (516) 979-2922

For specific information on your area, check with your local town or village planning board. The addresses of the planning boards for four of the eastern towns where greatest development is likely to occur are given below:

Town of Brookhaven
Planning Board
475 E. Main Street
Patchogue, NY 11772
Phone: (516) 475-5500

Town of East Hampton
Town Planner
159 Pantigo Road
East Hampton, NY 11937
Phone: (516) 267-8442

Town of Islip
Department of Planning
655 Main Street
Islip, NY 11751
Phone: (516) 224-5450

Southampton Town Planning Board
Town Hall Building
116 Hampton Road
Southampton, NY 11968
Phone: (516) 283-6000

Sewage and waste

For information, call or write:

Suffolk County Department of Health Services
General Engineering Unit
Riverhead, NY 11901
Phone: (516) 727-4700

Nassau County Department of Environmental Quality
240 Old Country Road
Mineola, NY 11501
Phone: (516) 535-3690

Vegetation

The soils and growing requirements of plants on barrier islands are different from conditions on mainland shores. For information regarding the planting of beach grass or other types of vegetation you can consult:

Suffolk County Soil and Water Conservation District Office
127 East Main Street
Riverhead, NY 11901
Phone: (516) 727-2315

You may also request information from the U.S. Department of Agriculture Soil Conservation Service at the same address as above.

Water resources

For information, call or write:

U.S. Geological Survey
Water Resources Division
5 Aerial Way
Syosset, NY 11591
Phone: (516) 938-8830

Suffolk County Department of Health Services
Water Quality Unit
65 Jetson Lane
Hauppauge, NY 11787
Phone: (516) 234-2622

Nassau County Department of Environmental Quality
240 Old Country Road
Mineola, NY 11501
Phone: (516) 535-3690

Suffolk County Water Authority
Sunrise Highway at Pond Road
Oakdale, NY 11796
Phone: (516) 589-5200

Weather service

Instant up-to-date forecasts can be had by dialing 936-1212 in Nassau and Suffolk counties. General information can be obtained from:

National Weather Service
Eastern Region
585 Stewart Avenue
Garden City, NY 11530
Phone: (516) 248-2101

Hurricane information is available from:

National Oceanic and Atmospheric Administration
Office of Coastal Zone Management
3300 Whitehaven Street, N.W.
Washington, DC 20235
Phone: (202) 634-6791

Appendix C. Useful references

The following publications are listed by subject; subjects are arranged in the approximate order that they appear in this book. A brief description of each reference is provided, and sources are included for those readers who would like to obtain more information on a particular subject. Most of the references listed are either low in cost or free. We encourage the reader to take advantage of these informative publications.

History

1. *Geomorphology of the South Shore of Long Island, New York*, by N. E. Taney, 1961. *Beach Erosion Board Technical Memorandum No. 128*, 28 pp. Discussion of the geology and historical shoreline change on the south shore of Long Island. Presents data relating to rates of shoreline retreat from old surveys of the high-water line. Available from the National Technical Information Service, Operations Division, 5285 Post Royal Road, Springfield, VA 22151.

Aerial photographs

2. *Aerial Photographic Resources of Nassau and Suffolk Counties, Long Island, New York*, by Thore Omholt, 1974, 16 pp. A catalogue of sources of aerial photographs giving scale, date, coverage of photographs, and address where they can be obtained. Available as technical report number 0029 from New York Ocean Science Laboratory, Montauk, NY 11954 or from local libraries.

Hurricanes

3. *Early American Hurricanes, 1492–1870*, by D. M. Ludlum, 1963. Informative and entertaining descriptions of storms affecting the Atlantic and Gulf coasts. Storm accounts, which appear in chronological order, provide insight into the frequency, intensity, and destructive potential of hurricanes. Published by the American Meteorological Society, Boston, MA. Available in public and university libraries.

4. *Atlantic Hurricanes*, by G. E. Dunn and B. I. Miller, 1960. Discusses at length hurricanes and associated phenomena such as storm surge, wind, and sea action. Includes a detailed account of Hurricane Hazel, 1954, and suggestions for pre- and posthurricane procedures. Published by the Louisiana State University Press, Baton Rouge, LA. Available in public and college libraries.

5. *Hurricane Information and Atlantic Tracking Chart*, by The National Oceanic and Atmospheric Administration, 1974. A brochure which describes hurricanes, defines terms, and lists hurricane safety rules. Outlines method of tracing hurricanes and provides a tracking map. Available from the Superin-

tendent of Documents, U.S. Government Printing Office, Washington, DC 20402.

6. *Bibliography on Hurricanes and Severe Storms of the Coastal Plains Region and Supplement*, by the Coastal Plains Center for Marine Development Services, 1970 and 1972. A list of references that provides a good starting point for persons seeking detailed information on hurricanes and hurricane research. Available through large libraries.

7. *A Wind to Shake the World*, by Everett S. Allen, 1976, 370 pp. Nicely written story of the events of the 1938 hurricane. The author had just joined the staff of the New Bedford *Standard-Times* when the storm struck. The book gives numerous first-hand accounts not available in other books about the 1938 storm. Published by Little, Brown and Company, Boston, MA. Available in public libraries.

8. *Hurricane*, by Joe McCarthy, 1969, 168 pp. The story of how Long Island and Connecticut were affected by the 1938 hurricane. Numerous personal accounts of the destructive force of this storm. The story of the Burghards' escape in chapter 1 of the present book is based on an account given in *Hurricane*. Published by the American Heritage Press and available in public libraries.

9. *The Hurricane of 1938 on Eastern Long Island*, by Ernest S. Clowes, 1939, 67 pp. A brief report, published by Hampton Press, Bridgehampton, L.I., taken from personal accounts on the effects the 1938 hurricane had on Long Island from Westhampton to Montauk Point. Includes some weather data on the storm and a partial estimate of damage costs. Available in public libraries.

10. *Storm Surge*, by Arthur Pole and Celso Barrientos, 1976. Describes the causes of storm surge and documents the meteorological situation that occurred during a number of major coastal storms. Available as *MESA New York Bight Atlas Monograph No. 6* from New York Sea Grant Institute, 411 State Street, Albany, NY 11246.

11. *Hurricane*, a 27-minute film that shows hurricane tracking and warning methods, and emphasizes precautions that should be taken in the event of a hurricane. Available from Film Librarian, Public Relations and Advertising Department, Aetna Life and Casualty, 151 Farmington Avenue, Hartford, CT 06115.

12. *Hurricane Decision*. This 14-minute film points out the dangers involved in storm surge, wind, and inland flooding. Available from Motion Picture Service, Department of Commerce—NOAA, 12231 Wilkins Avenue, Rockville, MD 20852.

Barrier islands

13. *Barrier Islands and Beaches*, 1976. Proceedings of the May 1976 barrier islands workshop. A collection of technical papers prepared by scientists studying barrier islands. Pro-

vides an up-to-date, readable overview of barrier islands. Subjects from aesthetics to flood insurance are discussed by the experts. Topics include island ecosystems, ecology, geology, politics, and planning. The book is a good bibliographic source for those interested in studying barrier islands. Available from the Publications Department, Conservation Foundation, 1717 Massachusetts Avenue N.W., Washington, DC 20036. (Request the Foundation's free list of publications.)

14. *Barrier Island Formation*, by J. H. Hoyt, 1967. Published in the *Bulletin of the Geological Society of America*, vol. 78, pp. 1125–36. A technical paper in which Hoyt develops his theory of barrier island formation. Available in university libraries or through interlibrary loan.

15. *Coastal Geomorphology*, edited by D. R. Coates, 1973. Another collection of technical papers including R. Dolan's "Barrier Islands: Natural and Controlled." Interesting reading for anyone willing to cope with the jargon of coastal scientists. Published by the State University of New York, Binghamton, NY 13901. Available in university libraries.

16. *Barrier Island Handbook*, by Stephen P. Leatherman, 1979. A very easy-to-read paperback that contains many explanatory illustrations. It considers barrier island dynamics in much greater detail than chapter 2 of this book. Many of the examples used come from the Maryland or New England coast but are directly applicable to the Long Island shore.

This book is recommended for the reader looking for a deeper understanding of barrier dynamics explained in nontechnical language. Available from Stephen Leatherman, Department of Geography, Rm. 1113 Social Science Building, University of Maryland, College Park, MD 20742.

17. *Know Your Mud, Sand, and Water: A Practical Guide to Coastal Development*, by K. M. Jurgensen, 1976. A clearly and simply written pamphlet describing various island environments relative to development. Recommended to island dwellers. Available from UNC Sea Grant, 1235 Burlington Laboratories, North Carolina State University, Raleigh, NC 27607.

18. *Barrier Beaches of the East Coast*, by P. J. Godfrey, 1976. Published in *Oceanus*, vol. 19, no. 5, pp. 27–40. A clearly written paper describing beaches and associated barrier island environments as related through island processes. This issue of *Oceanus* was devoted entirely to estuaries. Available in university libraries.

19. *Coastal Processes and Beach Erosion*, Joseph Caldwell, 1967. U.S. Army Coastal Engineering Research Center reprint. Published in the *Journal of the Society of Civil Engineers*, vol. 53, no. 2, pp. 142–157. This short paper attempts to explain some of the technical aspects of coastal processes at an elementary level. It is a good starting place for the individual interested in the more quantitative aspects of beach

erosion. It is available from the U.S. Army Coastal Engineering Research Center, Kingman Building, Ft. Belvoir, VA 22060

20. *Barrier Islands*, by Robert Dolan, Bruce Hayden, and Harry Lins, 1980. *American Scientist*, vol. 68, Jan.–Feb., pp. 16–25. An easy-to-read article describing the way in which barrier islands operate, and stressing the dangers involved in building on barrier islands. Available in university libraries or through interlibrary loan.

21. *Geomorphic Analysis of South Shore Barriers, Long Island, New York, Phase I*, by Stephen Leatherman and Danielle Joneja, 1980. A summary of literature relating to barrier origin and migration followed by a discussion of south-shore inlet history. The article is accompanied by a map showing inlet locations and positions of washovers, and was a principal reference in preparation of the maps for this volume. The article also contains a lengthy bibliography. Available from Stephen Leatherman, Department of Geography, Rm. 1113 Social Science Building, University of Maryland, College Park, MD 20742.

22. *Barrier Island Genesis: Evidence From the Central Atlantic Shelf, Eastern U.S.A.*, by Donald J. P. Swift, 1975. Published in *Sedimentary Geology*, vol. 14, pp. 1–43. Written in a technical style, this article supports the idea of barriers originating as a result of shoreface detachment (proposed

earlier by Hoyt, McGee, and Ganong). It brings into play evidence of prehistoric barrier retreat on the continental shelf and supports the concept of an offshore source for barrier sand. Available in university libraries or through interlibrary loan.

Geology

23. *Geomorphology, Shallow Subbottom Structure and Sediments of the Atlantic Inner Continental Shelf Off Long Island, New York*, by S. Jeffress Williams, 1976. U.S. Army Coastal Engineering Research Center, Technical Paper No. 76-2, 123 pp. A discussion of the nature of the layers of sediment that lie offshore from Long Island. It provides baseline information that will be used to delineate source areas for beach replenishment projects. Limited free copies are available from the U.S. Army Coastal Engineering Research Center, Kingman Building, Ft. Belvoir, VA 22060. Additional copies are available for a small fee from the National Technical Information Service, Operations Division, 5285 Port Royal Road, Springfield, VA 22151.

24. *Geomorphology and Sediments of the Inner New York Bight Continental Shelf*, by S. J. Williams and D. B. Duane, 1974. U.S. Army Corps of Engineers Coastal Engineering Research Center, Technical Memorandum no. 45. Similar in nature to the 1976 report of Williams (reference 23 above). Available from the National Technical Information Service,

Operations Division, 5285 Port Royal Road, Springfield, VA 22151.

25. *Evidence of Shoreface Retreat and In Place "Drowning" During Holocene Submergence of Barriers, Shelf Off Fire Island, New York*, by John E. Sanders and Naresh Kumar, 1975. *Journal of Sedimentary Petrology*, vol. 86, p. 65–76. This article suggests that the south-shore barriers have moved landward in great jumps rather than by gradual retreat. As sea level rose ancient barriers are thought to have been submerged and new barriers formed landward of the old islands. Available in most college or university libraries.

26. *Soil Survey of Suffolk County New York*, 1975. A detailed two-part report on the soils of Long Island. The first part describes the use, management, and types of soils that occur on Long Island. The second part is a collection of photographic maps that show the distribution of each soil type. It is available at most libraries.

Beaches

27. *Waves and Beaches*, by Willard Bascom, 1964. Published by Anchor Books, Doubleday and Co., Garden City, NY 11530. An easy-to-read discussion of wave dynamics and coastal processes. Available in paperback from local bookstores.

28. *Beaches and Coasts*, Second edition, by C. A. M. King, 1972. Published by St. Martin's Press, Inc., 175 Fifth Avenue, New York, NY 10010. A classic treatment of beach and coastal processes, this is detailed but very tedious reading. Available in public libraries.

29. *Beach Processes and Sedimentation*, by Paul Komar, 1976. Published by Prentice Hall, Englewood Cliffs, NJ 07632. The most up-to-date technical explanations of beaches and beach processes. Recommended only to serious students of the beach. Available in university libraries or through interlibrary loan.

30. *Land Against the Sea*, by the U.S. Army Corps of Engineers, 1964. Readable introduction to coastal geology and shoreline processes. However, the authors' belief in the value of certain engineering methods is either outdated or unsubstantiated. Available as Miscellaneous Paper No. 4-64 from the U.S. Army Corps of Engineers, Coastal Engineering Research Center, Kingman Building, Ft. Belvoir, VA 22060.

31. *Atlantic Beaches*, by J. N. Leonard, 1972. Published as part of the American Wilderness Series by Time-Life Books, Rockefeller Center, New York, NY 10020. Presents the beauty of the beach in words and pictures. Available in university and public libraries.

32. *Beach Forms and Coastal Processes*, by Warren Yasso and Elliott Hartman, Jr., 1975. Presents a brief survey of natural processes and man-made beaches, dunes, and shoreline structures along the New Jersey–New York shore. Emphasizes

the point that coastal structures will compound cost of storm repair, and the futility of shore protection methods. Available as *MESA New York Bight Atlas Monograph No. 11* from New York Sea Grant Institute, 411 State Street, Albany, NY 11246.

33. *Beach: A River of Sand.* An excellent film that discusses coastal processes and the effects of man's interference with the natural system. Available from Film Rental Center, Syracuse University, 1455 East Colvin Street, Syracuse, NY 13210.

Recreation

34. *Recreation in the Coastal Zone*, 1975. A collection of papers presented at a symposium sponsored by the U.S. Department of the Interior, Bureau of Outdoor Recreation, Southeast Region. Outlines different views of recreation in the coastal zone, and the approaches taken by some states to recreation-related problems. The symposium was cosponsored by the Office of Coastal Zone Management. Available from that office, National Oceanic and Atmospheric Administration, 3300 Whitehaven Street N.W., Washington, DC 20235.

Erosion

35. *Sea Level Rise as a Cause of Shore Erosion*, by Per Bruun, 1962. *Journal of Waterways and Harbors Division, Proceedings of the American Society of Civil Engineers*, vol. 88, pp. 117–31. This technical article offers a widely accepted explanation for the erosion of shorelines under the condition of rising sea level. Although we do not see it as applicable to the south shore of Long Island, it has been widely quoted as an explanation for erosion all along the eastern seaboard. Available in some university libraries or through interlibrary loan.

36. *Probable Causes of Shoreline Recession and Advance on the South Shore of Eastern Long Island*, by C. L. McCormick, 1973. Pages 61–71 in *Coastal Geomorphology*, edited by D. R. Coates. Publications in Geomorphology, State University of New York at Binghamton. This article suggests that the increase in erosion after 1930 at Westhampton Beach was caused by the opening and maintenance of Moriches and Shinnecock inlets. Available in university libraries.

37. *Interim Report on Task 7.2, Flood Plain Management, Beach Erosion and Hurricane Damage Reduction South Shore of Long Island*, 1976, 67 pp. A report summarizing information contained in several reports prepared by the U.S. Army Corps of Engineers, New York District. Surveys the problems and alternative solutions available for each segment of the south shore. It is available from the New York State Department of Environmental Conservation, Office of Program Development, Planning and Research, 50 Wolf Road, Albany, NY 12233.

Shoreline engineering

38. *Shore Protection Guidelines*, by the U.S. Army Corps of Engineers, 1971. Summary of the effects of waves, tides, and winds on beaches and engineering structures used for beach stabilization. Available free from the Department of the Army, Corps of Engineers, Washington, DC 20318.

39. *Low Cost Shore Protection*, by the U.S. Army Corps of Engineers, 1982. Four reports in a set written for the layman are listed under this title. They include the introductory report, a property-owner's guide, a guide for local government officials, and a guide for engineers and contractors. The reports are a summary of the Shoreline Erosion Control Demonstration Program and suggest a wide range of engineering devices and techniques to stabilize shorelines, including beach nourishment and vegetation. In adopting these approaches, you should keep in mind that they are short-term measures and may have unwanted side effects. The reports are available from John G. Housley, Section 54 Program, U.S. Army Corps of Engineers, USACE (DAEN-CWP-F), Washington, DC 20314.

40. *Beach Behavior in the Vicinity of Groins*, by Craig H. Everts, 1979. Pages 853–67 in *Proceedings of the Specialty Conference on Coastal Structures 79, Alexandria, Virginia*. An interesting description of the effects of two groin fields in New Jersey. Concludes that groins deflect the movement of sand seaward, causing erosion in the downdrift shadow area. Suggests that groins have negative downdrift effect even if groin compartments are filled with sand. Available from the U.S. Army Coastal Engineering Research Center, Kingman Building, Ft. Belvoir, VA 22060 as reprint 79-3.

41. *Beach Changes at Westhampton Beach, New York, 1962–1973*, by Allan E. DeWall, 1979. *U.S. Army Coastal Engineering Research Center Miscellaneous Report No. 79-5.* This report presents data gathered over an 11-year period showing in some detail how the groins at Westhampton Beach altered the natural shoreline. The conclusions are not earthshaking: groins trap sand, and downdrift areas erode. Limited free distribution is made by the U.S. Army Coastal Engineering Research Center, Kingman Building, Ft. Belvoir, VA 22060. Additional copies are available for a small fee from the National Technical Information Service, Operations Division, 5285 Port Royal Road, Springfield, VA 22161.

42. *The Atlantic Coast of Long Island*, by Frank L. Panuzio, 1969. Pages 1222–41 in *Proceedings of the Conference on Coastal Engineering*, vol. 11. A summary article describing the littoral processes on the south shore and the kinds of structural solutions proposed by the U.S. Army Corps of Engineers to deal with problem areas. It also discusses the effects of structural solutions implemented by 1969. This is a good and easy-to-read summary. Available in university libraries.

43. *Shore Protection Manual*, by the U.S. Army Corps of Engineers, 1973. The "bible" of shoreline engineering. Published in three volumes. Request publication 08-0-22-00077 from the Superintendent of Documents, U.S. Government Printing Office, Washington, DC 20402.

44. *Help Yourself*, by the U.S. Army Corps of Engineers. Brochure addressing the erosion problems in the Great Lakes region. May be of interest to barrier island residents because it outlines shoreline processes and illustrates a variety of shoreline-engineering devices used to combat erosion. Free from the U.S. Army Corps of Engineers, North Central Division, 219 South Dearborn Street, Chicago, IL 60604.

45. *Publications List, Coastal Engineering Research Center (CERC) and Beach Erosion Board (BEB)*, by the U.S. Army Corps of Engineers, 1976. A list of published research by the U.S. Army Corps of Engineers. Free from the U.S. Army Corps of Engineers, Coastal Engineering Research Center, Kingman Building, Ft. Belvoir, VA 22060.

Hazards

46. *Natural Hazard Management in Coastal Areas*, by G. F. White and others, 1976. A summary of coastal hazards along the entire U.S. coast. Discusses adjustments to hazards, and covers hazard-related federal policy and programs. Summarizes hazard management and coastal land planning programs in each state. Appendices include a directory of agencies, an annotated bibliography, and information on hurricanes. An invaluable reference, recommended to developers, planners, and managers. Available from the Office of Coastal Zone Management, National Oceanic and Atmospheric Administration, 3300 Whitehaven Street N.W., Washington, DC 20235.

47. *Guidelines for Identifying Coastal Hazard Zones*, by the U.S. Army Corps of Engineers, 1975. Report outlining high-hazard zones, with emphasis on "coastal special flood-hazard areas" (coastal floodplains subject to inundation by hurricane surge with a one-percent chance of occurring in any given year). Provides technical guidelines for conducting uniform flood insurance studies, and outlines methods of obtaining 100-year storm-surge elevations. Recommended to island planners. Available from the Galveston District, U.S. Army Corps of Engineers, Galveston, TX 77553.

48. *The Homeport Story*. The fictional story of a coastal community preparing for a hurricane disaster. It is intended to act as a model for communities that wish to develop a preparedness plan. Available from the National Oceanic and Atmospheric Administration, Rockville, MD 20852 or from the U.S. Government Printing Office, Washington, DC 20402. If ordering from the Government Printing Office, refer to stock no. 0317-0046.

Flood Insurance

49. *Questions and Answers on the National Flood Insurance Program*, by the Federal Emergency Management Agency (FEMA), 1983. This pamphlet explains basics of flood insurance, and provides addresses of FEMA offices. Free from Federal Emergency Management Agency, Washington, DC 20472.

50. *Coastal Flood Hazards and the National Flood Insurance Program*, by H. Crane Miller, 1977, 50 pp. This publication describes in detail the nature of flood hazards in the coastal zone and the basic features of the flood insurance program. The third and most interesting portion of this article is the attempt to assess the impact that this program has had on those interested in building in the coastal zone. Available from National Flood Insurance Program, Federal Emergency Management Agency, Washington, DC 20472.

51. *Guide for Ordinance Development, No. 1e.* A guidebook designed for use by community officials in preparing flood-plain management measures which satisfy the minimum standards of the national flood-insurance program. It organizes the program's standards into a simple ordinance, and provides an explanatory narrative. Available from Federal Emergency Management Agency, Washington, DC 20472.

52. *Entering the Regular Program, No. 3.* Intended for use by community officials during the period when a community enters the regular flood insurance program. Explains the responsibilities of the FEMA and the local community that must be met to enter the regular program. It includes a timetable that should be followed. Available from Federal Emergency Management Agency, Washington, DC 20472.

Water supply and sewage

53. *Standards for Sewage and Waste Disposal Systems, Design of Residential Subsurface Sewage Disposal Facilities, Part 1.* Describes in some detail the minimum requirements for installation of a private sewage system complete with specifications. This publication available from Suffolk County Department of Health Services, General Engineering Unit, Suffolk County Center, Riverhead, NY 11901, phone: (516) 727-4700.

54. *Suffolk County Department of Health Services Procedure and Requirements for Residential Construction.* A one-page description of the procedure to be followed for approval to construct a water supply facility and a sewage disposal system. Available from Suffolk County Department of Health Services, General Engineering Unit, Suffolk County Center, Riverhead, NY 11901, phone: (516) 727-4700.

55. *Individual Water Supply Systems, Preliminary Bulletin.* A one-page description of requirements for private wells. Available from Suffolk County Department of Health Services,

General Engineering Unit, Suffolk County Center, River-head, NY 11901, phone: (516) 727-4700.

56. *Comprehensive Public Water Supply Study Suffolk County, New York*, by Holzmacher, McLandon and Murrell, Consulting Engineers, 1970. CPWS-24, V 11, 373 pp. Technical and, as the title implies, comprehensive. Available from Suffolk County Department of Health Services, General Engineering Unit, Suffolk County Center, Riverhead, NY 11901.

57. *Standards for Sewage Waste Disposal Systems, Design of Nonresidential Subsurface Sewage Disposal Facilities, Part II*. Describes in some detail the requirements for installation of a disposal system for small commercial establishments. Available from Suffolk County Department of Health Services, General Engineering Unit, Suffolk County Center, Riverhead, NY 11901.

58. *Sewage Disposal and Other Environmental Factors, Nassau County Department of Health Manual*. A manual prepared for engineers, architects, and surveyors for design and installation of on-site sewage disposal facilities in developments. This may not be too helpful to citizens of Nassau County as most of the county has access to a sewer system. Available from Nassau County Department of Environmental Quality, 240 Old Country Road, Mineola, NY 11501, phone: (516) 535-3690.

59. *Long Island Comprehensive Waste Treatment Plan (208 Plan)*, 1978. A two-volume report covering all aspects of water supply and waste treatment on Long Island. The first volume contains a general statement on the nature of the problems faced by Long Island, and gives available planning options. The second volume summarizes documentation of the types and amounts of pollution on the island. It contains a complete bibliography on the subjects treated and is available in the reference section of most libraries on Long Island.

60. *Hydrogeology of Suffolk County, Long Island, New York*, by H. M. Jensen and J. Soren, 1974. U.S. Geological Survey, Washington, DC 20244. A set of maps with attached explanations giving a concise view of the Suffolk County groundwater situation. Easy to read, with a minimum of technical jargon. Available in university libraries.

Vegetation

61. *Ecological Studies of the Sunken Forest Fire Island National Seashore, New York*, by Henry Warren Art, 1976. National Park Service Scientific Monograph Series, No. 7, 237 pp. A detailed discussion of the plant distribution, primary production, and nutrient relationships of the Sunken Forest area. It contains chapters on the more general topics of geology and climate and vegetational patterns of other Atlantic barrier islands. It is available from the Superintendent of Documents, Government Printing Office, Washington, DC 20402.

62. *Environmental Inventory of the Fire Island National Seashore and William Floyd Estate, Suffolk County, New York*, 1975. A report that summarizes virtually all available information pertaining to the physical and biological environments within the boundaries of the national seashore up to 1975. It is dull, unimaginative reading but filled with information and it has an exhaustive bibliography. The report was produced for the National Park Service by the private consulting firm of Jack McCormick and Associates and submitted to the National Park Service, U.S. Department of Interior, Denver Service Center, Denver, Colorado. Depending on supply, copies might be obtained from the latter source or directly from the Fire Island National Seashore, 120 Laurel Avenue, Patchogue, NY 11772, phone: (516) 289-4810.

63. *How to Build and Save Beaches and Dunes*, by John A. Jagschitz and Robert C. Wakefield, 1971. *University of Rhode Island Marine Leaflet Series No. 4*, 12 pp. Outlines how to build dunes using brush, snow fencing, or American beach grass. Most of the discussion pertains to the planting and caring of beach grass. Available from Rhode Island Agricultural Experiment Station, Woodward Hall, University of Rhode Island, Kingston, RI 02881.

64. *Discovering Fire Island: The Young Naturalist's Guide to the World of the Barrier Beach*, 1978. An instructive guide to the vegetation and animals that inhabit the barrier beach. Intended for young readers. Available from the Fire Island National Seashore, 120 Laurel Avenue, Patchogue, NY 11772, phone: (516) 289-4810.

Planning and management

65. *Shore Management Guidelines* and *Shore Protection Guidelines* are two publications produced by the U.S. Army Corps of Engineers; they are designed to instruct the public on the use of the shore and the use of structural solutions in solving erosion problems. Both are available from the Department of the Army, Corps of Engineers, Washington, DC 20314.

66. *Guidelines for Long Island Coastal Management*, 1973. This report by the Regional Marine Resources Council is an expression of planning guidelines applicable to four areas of concern in the coastal environment: (1) coast stabilization and protection, (2) dredging and dredge spoil disposal, (3) wetlands management, and (4) water supply and waste disposal. It also contains an annotated bibliography and selected reference list for the topics listed above. Available from the Regional Marine Resources Council, Veterans Memorial Highway, Hauppauge, NY 11787.

67. *Master Plan, Fire Island National Seashore*, 1975. A management plan that takes into consideration the dynamic nature of barrier islands. Management is intended to remain flexible in order to accommodate whatever changes the future offers. Limited copies are available from the National Park Service, 120 Laurel Avenue, Patchogue, NY 11772.

68. *Politics of Shore Erosion: Westhampton Beach*, by Joseph M. Heikoff, 1976. Ann Arbor Science Publishers, Inc., Ann Arbor, MI 48106, 173 pp. A detailed discussion of the events that led to the construction of the groin field at Westhampton Beach. Documents the failure of public agencies to cooperate and the problems generated when decisions were made to proceed with construction even though it violated recommended engineering practice. This is an eye-opening account of a situation where the failure of public agencies to cooperate jeopardized the success of a project. Available in university libraries.

69. *The Water's Edge: Critical Problems of the Coastal Zone*, edited by B. H. Ketchum, 1972. Published by the M.I.T. Press, Cambridge, MA 02139. The best available scientific summary of coastal zone problems. Available in university libraries.

70. *Design with Nature*, by Ian McHarg, 1969. Published by Doubleday and Company, Inc., Garden City, NY 11530. A now-classic text on the environment. Stresses that when man interacts with nature, he must recognize its processes and governing laws, and realize that it both presents opportunities for, and requires limitations on, human use. Available in university libraries.

71. *Coastal Ecosystems, Ecological Considerations for Management of the Coastal Zone*, by John Clark, 1974. A clearly written, well-illustrated book on the applications of ecologic principles to the major coastal zone environments. Available from the Publications Department, The Conservation Foundation, 1717 Massachusetts Avenue N.W., Washington, DC 20036.

72. *Who's Minding the Shore*, by the Natural Resources Defense Council, Inc., 1976. A guide to public participation in the coastal zone management process. Defines coastal ecosystems and outlines the Coastal Zone Management Act, coastal development issues, and means of citizen participation in the coastal zone management process. Lists sources of additional information. Available from the Office of Coastal Zone Management, National Oceanic and Atmospheric Administration, 3300 Whitehaven Street N.W., Washington, DC 20235.

73. *Coastal Development and Areas of Environmental Concern*, edited by Simon Baker, 1975. Proceedings of a symposium held at East Carolina University in Greenville, North Carolina, in March 1975. The collection of papers outlines sensitive environments, types of sites to be preserved, and implementations of the North Carolina Coastal Area Management Act of 1974. Interesting papers on sand dunes, salt marshes, archaeological sites, and historic sites. Especially recommended is the paper entitled "Scenery for Sale" (pp. 28–41) by A. Conrad Neumann, which summarizes the natural, economic, and social pressures exerted on barrier islands. Available from UNC Sea Grant, 1235 Burlington

Laboratories, North Carolina State University, Raleigh, NC 27607.

74. *Termination II: How the National Park Service Annulled its "Commitment" to a Beach Erosion Control Policy at the Cape Hatteras National Seashore*, by R. D. Behn and M. A. Clark, 1976. A lesson in modern history about a very significant change in the policy of the National Park Service. The study traces the agency's initial commitment to shoreline engineering on the Outer Banks of North Carolina, the futile spending of millions of dollars on "protective" projects, and the gradual abandonment of the commitment in the mid-1970s. Interesting reading. Available from the Center for Policy Analysis, Institute of Policy Sciences and Public Affairs, Duke University, Durham, NC 27706.

75. *The Fiscal Impact of Residential and Commercial Development: A Case Study*, by T. Muller and G. Dawson, 1972. A classic study which demonstrates that development may ultimately increase, rather than decrease, community taxes. Available from the Publications Office, the Urban Institute, 2100 M Street N.W., Washington, DC 20037. Refer to URI-22000 when ordering.

76. *Report of the Conference on Marine Resources of the Coastal Plains States*, 1974. Collection of papers presented at a meeting in Wilmington, North Carolina. Topics include seabed mineral resources, sport fishing, recreation and tour-

ism, and coastal zone planning. Of special interest is a paper entitled "Responsible Development and Reasonable Conservation," by David Stick. Sponsored and published by the Coastal Plains Center for Marine Development Services, 1518 Harbour Drive, Wilmington, NC 28401.

Directories

77. *Long Island Environmental Directory*, 1979. A directory of government and private conservation groups on Long Island. This directory is available from the Marine Environmental Council of Long Island, Inc., 2348A Maple Street, Seaford, NY 11783. A small donation is requested for the directory.

78. *Environmental Directory*. Available from the New York State Department of Environmental Conservation, 50 Wolf Road, Albany, NY 12233.

79. *Environmental Assistance Directory*. This directory lists information sources by topic and is available from the Council on Environmental Quality, H. Lee Dennison Building, Hauppauge, NY 11787.

Individual barrier islands

80. *Staten Island, New York, Fort Wadsworth to Arthur Kill, Beach Erosion Control and Hurricane Protection Project, General Design Memorandum No. 1, Fort Wadsworth to Great Kills Park*, 1976. One of a series of reports produced by the New York District of the U.S. Army Corps of Engi-

neers. These reports present a reasonably comprehensive survey of information available on the physical environment. Topics covered include littoral forces, shore history, and storm history, as well as identification of existing shoreline problems and proposed solutions. They are not always easy to acquire but are available in some libraries and as supplies permit from the U.S. Army Corps of Engineers, New York District Office, 26 Federal Plaza, New York, NY 10007. Titles of other similar reports are given in references 81–85 below.

81. *Atlantic Coast of New York City from Rockaway Inlet to Norton Point, New York (Coney Island Area), Cooperative Beach Erosion Control and Interim Hurricane Study*, 1973. For information on how to acquire this report see 80 above.

82. *Beach Erosion Control and Interim Hurricane Study of the Atlantic Coast of Long Island, New York, Jones Inlet to East Rockaway Inlet*, 1973. For information on how to acquire this report, see 80 above.

83. *Atlantic Coast of Long Island, New York, Fire Island Inlet and Shore Westerly to Jones Inlet, Beach Erosion Control Report on Cooperative Study*, 1955. For information on how to acquire this report, see 80 above.

84. *Atlantic Coast of Long Island, New York, Fire Island Inlet to Montauk Point, Cooperative Beach Erosion Control and Interim Hurricane Study*, 1958. For information on how to acquire this report, see 80 above.

85. *Fire Island Inlet to Montauk Point, Long Island, New York, Beach Erosion and Hurricane Project, Supplement No. 1 to General Design Memorandum No. 1, Moriches to Shinnecock Beach*, 1969. This report basically covers the progress of work at Westhampton Beach. For information on how to acquire this report, see 80 above.

86. *Oak Beach Erosion Study*, no date. A report submitted to the Town of Babylon by Greenman-Pedersen, Associates, P.C. This brief report describes the nature of the erosion problem at Oak Beach and suggests causes and possible solutions for the problems. A limited number of copies of this report might be available from the Commissioner of Environmental Control, Town of Babylon, 200 East Sunrise Highway, Lindenhurst, NY 11757.

Building or improving a home

Both current and prospective owners and builders of homes in hurricane-prone areas should supplement the information and advice provided in this book with that offered in the following references. These excellent references contain sound, useful information that should help the residents of such areas to minimize the losses caused by extreme wind or rising water. Many of these publications are free. The government publications are paid for by your taxes, so why not use them?

87. *Coastal Design: A Manual for Planners, Developers, and Home Owners*, by Orrin H. Pilkey, Jr., Orrin H. Pilkey, Sr.,

Walter D. Pilkey, and William J. Neal, 1983. A detailed construction guide expanding on the information given in chapter 6 of the present book. Chapters include discussions of shoreline types, individual residence construction, making older structures storm-worthy, high-rise buildings, mobile homes, coastal regulations, and the future of the coastal zone. Published by Van Nostrand Reinhold Co., New York, NY.

88. *Design and Construction Manual for Residential Buildings in Coastal High Hazard Areas*, prepared by Dames and Moore for the Department of Housing and Urban Development on behalf of the Federal Emergency Management Agency, Federal Insurance Administration, 1981. A guide to the coastal environment with recommendations on site and structure design relative to the National Flood Insurance Program. The report includes design considerations, examples, construction costs, and appendices on design tables, bracing, design worksheets, wood preservatives, and a listing of useful references. The manual is available from the Superintendent of Documents, U.S. Government Printing Office, Washington, DC 20402 (publication number 722-967/545) or contact an office of the Federal Emergency Management Agency.

89. *Elevated Residential Structures, Reducing Flood Damage through Building Design: A Guide Manual*, prepared by the Federal Insurance Administration, 1976. An excellent outline of the flood threat and necessity for proper planning and construction. Illustrates construction techniques and includes glossary, references, and worksheets for estimating building costs. Order publication 0-222-193 from the Superintendent of Documents, U.S. Government Printing Office, Washington, DC 20402, or contact an office of the Federal Emergency Management Agency.

90. *Wind-Resistant Design Concepts for Residences*, by Delbart B. Ward. Displays with vivid sketches and illustrations construction problems and methods of tying structures down to the ground. Considerable text and excellent illustrations devoted to methods of strengthening residences. Offers recommendations for relatively inexpensive modifications that will increase the safety of residences subject to severe winds. Chapter 8, "How to Calculate Wind Forces and Design Wind-Resistant Structures," should be of particular interest to the designer. Available as TR-83 from the Civil Defense Preparedness Agency, Department of Defense, The Pentagon, Washington, DC 20301; or the Civil Defense Preparedness Agency, 2800 Eastern Boulevard, Baltimore, MD 21220.

91. *Interim Guidelines for Building Occupant Protection from Tornadoes and Extreme Winds*, TR-83A, and *Tornado Protection—Selecting and Designing Safe Areas in Buildings*, TR-83B. These are supplement publications to the above reference and are available from the same address.

92. *Standard Building Code* (1979, previously known as the *Southern Standard Building Code* and still frequently re-

ferred to by that name). Available from Southern Building Code, Congress International, Inc., 900 Montclair Road, Birmingham, AL 35213.

93. *Structural Failures: Modes, Causes, Responsibilities*, 1973. See especially the chapter entitled "Failure of Structures due to Extreme Winds," pp. 49–77. Available from the Research Council on Performance of Structures, American Society of Civil Engineers, 345 East 47th Street, New York, NY 10017.

94. *Hurricane-Resistant Construction for Homes*, by T. L. Walton, Jr., 1976. An excellent booklet produced for residents of Florida, but equally useful to those of the New York coast. A good summary of hurricanes, storm surge, damage assessment, and guidelines for hurricane-resistant construction. The booklet gives technical concepts on probability and its implications for home design in hazard areas. There is also a brief summary of federal and local guidelines. Available from Florida Sea Grant Publications, Florida Cooperative Extension Service, Marine Advisory Program, Coastal Engineering Laboratory, University of Florida, Gainesville, FL 32611.

95. *Guidelines for Beachfront Construction with Special Reference to the Coastal Construction Setback Line*, by C. A. Collier and others, 1977, Report No. 20. Available from Florida Sea Grant Publications, Florida Cooperative Extension Service, Marine Advisory Program, Coastal Engineering Laboratory, University of Florida, Gainesville, FL 32611.

96. *Houses Can Resist Hurricanes*, by the U.S. Forest Service, 1965. An excellent paper with numerous details on construction in general. Pole-house construction is treated in particular detail (pp. 28–45). Available as *Research Paper FPL 33* from Forest Products Laboratory, Forest Service, U.S. Department of Agriculture, P.O. Box 5130, Madison, WI 53705.

97. *Pole House Construction* and *Pole Building Design*. Available from the American Wood Preservers Institute, 1651 Old Meadows Road, McLean, VA 22101.

98. *Standard Details for One-Story Concrete Block Residences*, by the Masonry Institute of America. Contains nine fold-out drawings that illustrate the details of constructing a concrete-block house. It presents principles of reinforcement and good connections aimed at design for seismic zones, but these apply to design in hurricane zones as well. Written for both layman and designer. Available as Publication 701 from Masonry Institute of America, 2550 Beverly Boulevard, Los Angeles, CA 90057.

99. *Masonry Design Manual*, by the Masonry Institute of America. A very thorough manual that covers all types of masonry including brick, concrete block, glazed stuctural units, stone, and veneer. Very comprehensive and well presented. Probably of more interest to the designer than to the layman. Available as Publication 601 from the Masonry Institute of America, 2550 Beverly Boulevard, Los Angeles, CA 90057.

Appendix D. Field trip guides

Trip 1. The south shore of eastern Long Island

This field trip guide begins in the parking area at Montauk Point Lighthouse and follows the south shore 68.3 miles west to the washover zone at Westhampton Beach. Driving time is about 2 hours (this does not include time spent at each stop). There are seven recommended stops along this route. The entire trip with stops, walking, and time out for lunch is about six hours, but this time can be considerably shortened or lengthened depending on your pace.

This trip should *not* be taken during July or August because most of the parking areas require village or town parking stickers.

0.0 **Stop 1. Montauk Lighthouse**. This is one of the most scenic spots on eastern Long Island. It is best viewed early in the morning when the sun strikes the eroding bluffs below the lighthouse. The site for the light was chosen in 1792, supposedly by George Washington. It was situated well back from the edge of the bluff to protect it from erosion. It was certainly a good idea since the bluff has retreated rapidly since the light was built (see fig. 4.2).

You can follow a path to the beach from the concession building across from the parking area. The beach is composed of boulders, which form a kind of natural armor that slows, but does not stop, the erosion of this area. The boulders were derived from erosion of the bluffs that the lighthouse crowns. They were brought in by glacial ice that deposited the material that forms Montauk Point. Too big to be moved very far from the point where they have eroded, the boulders have collected here, forming what geologists refer to as a lag deposit. (It lags behind the more mobile sand and clay constituents that have been eroded from the bluffs.) Sand is washed along the shore to nourish beaches to the west, while the finest clay and silt are lost offshore, presumably to come to rest in some deep spot on the seabed where currents are weak enough to allow this material to remain.

There is no question that erosion of these bluffs provides a substantial amount of sand needed to maintain the sandy beaches further west. You might note that the base of the bluff has been artificially armored with large blocks of stone to stop erosion of the bluff and preserve the light. What would happen if this technique were used to protect the bluffs all the way to Napeague Harbor? The answer, of course, is that beaches to the west would erode. The south shore operates as an integrated system. Changes in one part of that system will result in additional changes somewhere else in the system.

Exit left from parking area and follow Route 27.

2.3 Scenic overlook on left.

4.2 Ditch Plains Road on left.

5.4 Enter village of Montauk.

6.3 **Turn left** as you leave Montauk Village onto Old Montauk Highway. Here the road follows an old bluff (on left) cut into the glacial deposits. The bluff is now separated from the sea by a dune area that narrows westward toward Gurney's Inn.

8.6 Gurney's Inn on left.

9.8 Hither Hills State Park.

10.5 **Bear left** on Route 27. Note the unvegetated sandy area on the hill in front of your vehicle at this intersection. This is the front portion of one of the so-called walking dunes in this area. Except for the bare area, this particular dune is almost entirely stabilized by vegetation and attains an elevation of about 80 feet.

11.1 **Turn right** on Napeague Harbor Road; the Hither Hills Racquet Club is on the northwest corner of this turn.

11.8 *Stop 2. End of Napeague Meadow Road.* Park your vehicle in the small turnaround area at the end of this road and follow the unmarked sandy path leaving the east side of the turnaround area. This path will lead you up the west flank of the most active of the three walking dunes in this area (see fig. 4.5). Note that the dune seems to be moving across a surface as flat as the top of a billiard table. That is because the northwest winds that sweep the sand from the back (northwest) side of the dune to the front (southeast) side can only remove sand down to the level of the water table. At this point the grains stick to each other because they are wet and unable to be moved by the wind.

Note the tree trunks projecting from the back side of the dune. They are the remnants of a forest that was buried as the dune shifted to the southeast. Now they are being uncovered on the back side of the dune.

If you walk to the active front slope of the dune and look to the southeast, you can see the vegetated form of two more of these large parabolic-shaped dunes. Both are almost entirely stabilized by vegetation.

A student in one of my classes has suggested a reasonable explanation for the origin of these dunes. He suggested that the westward-flowing littoral drift brought sand eroded from the bluffs along the *north* side of the south fork of Long Island into this area. As the shore built northward toward Gardiner's Island the northwest winds swept the beach sand into dunes leaving a trail of three large parabolic dunes that exist today. If the degree of stabilization of the dunes is a measure of their relative age this theory seems quite reasonable. The most southerly dune is the most highly stabilized in the sequence, and the northernmost dune is the most active.

Return to vehicle and drive back to Route 27 on Napeague Harbor Road.

12.6 **Turn right** on Route 27.

13.7 **Turn right** on Napeague Meadow Road. This turn is located opposite the big twin antenna towers on the right side of Route 27. Watch carefully, it's an easy turn to miss.

15.2 **Turn left** onto Cranberry Hole Road.

15.4 *Stop 3. Promised Land fish plant.* North of your vehicle is the old fishmeal plant at Promised Land. Operation of the plant was discontinued some years ago when the menhaden catch dropped off. South of the road is a low but very obvious dune ridge that extends west. This is the dune ridge that built on top of a spit that grew westward from Promised Land island until it reached the glacial deposits you can see to the west. The position of this dune ridge and former spit is shown on figure 4.5.

To the east you can see the walking dunes and the glacial ridge that forms most of the south fork of Long Island. To the west more glacial deposits are visible. The land between is the result of deposition of beach and dune sediment from the littoral drift. The area where you stand was once shallow sea bottom. It was brought above sea level as sand washed in from an island that existed a short distance northeast of the fish plant. At the same time a spit built westward from Hither Hills State Park and eventually attached to the dune line visible at this locality. The spit from Hither Hills and the spit from Promised Land formed a narrow land bridge that connected Hither Hills with the rest of Long Island.

After the connection was complete, sand in the littoral drift straightened the shore by filling in the Napeague area on both the north and south sides of the original spit.

Turn your vehicle around and return to Route 27 the same way you came into the old fish plant.

15.6 **Turn right** on Napeague Meadow Road. Note as you travel down this road that there is a nice example of floral succession as salt-marsh vegetation fills in the shallow end of Napeague Harbor on the left. The salt marsh gradually gives way to salt-tolerant shrubs, and these are replaced in their turn by pine trees seen here on the right side of the road.

17.1 **Turn right** onto Route 27.

20.2 At this point Route 27 climbs the face of an old sea cliff that was cut into the glacial deposits before the littoral drift that bridged Napeague filled in this area.

20.8 **Turn left** onto Bluff Road, named for the old sea cliff that borders this road on the left. Note the wide expanse of dunes between this bluff and the present shoreline. The belt of dunes below the bluff narrows progressively to the west and disappears completely before East Hampton Village is reached.

21.7 East Hampton Town Marine Museum is on the left. This small museum, dealing with the history of local boats, is well worth the time it takes to stop and look around.

22.4 **Turn right** onto Indian Wells Road.

22.9 **Turn left** onto Route 27.

26.3 **Traffic light**. Travel through the village of East Hampton. At this traffic light Route 27 turns 90 degrees to the right. Do not follow Route 27; go straight ahead on Ocean Avenue.

27.1 ***Stop 4. Main beach at East Hampton***. The sandy beach at this stop cuts across the eroded terminus of old glacial outwash deposits. Walk west past the beach pavilion. Prior to 1978 there were almost no shore-hardening structures on this shore. Now the village beach is almost entirely lined with one type or another of shore-hardening device (fig. 4.7). I frequently refer to this section of beach as my coastal chamber of horrors. It begins with a stone revetment, followed by sheet-aluminum piling, and this in turn is followed by a massive wooden bulkhead. Beyond the wood is another stone revetment. At the end of the stone revetment is a small expanse of shore that has been left in a natural state. As long as it exists, it forms a very interesting comparison with the artificially protected portions of the shore. I measured the width of the beach at this point in the spring of 1982. It was 135 feet wide from the base of the bluff to the crest of the beach face, while the beach in front of the adjacent stone revetment measured slightly less than 40 feet.

In addition to this casual observation, it is clear that over the years the shore in this area has retreated upward of one mile. Despite the obvious continual trend of shoreline retreat along this section of beach, shore-front homeowners argue that the beaches are not retreating. To do so would be to admit that their shore-hardening structures would in the long run eliminate the beach. There is something terribly inconsistent about a group of people arguing on the one hand that the shore is not retreating and on the other spending millions of dollars to stop the retreat. Perhaps even more ironic is that the group that represents these shore-front interests is called the East Hampton Beach Preservation Society.

Recently the Town of East Hampton attempted to pass legislation that would prevent the further construction of these structures. In a heated debate over this issue in public hearings, the day was won by the homeowners, but not until they threatened to use their considerable wealth to tie the town up in court for years on constitutional issues. In the end it seemed to be the threat of prolonged court battles that settled the issue rather than concern over the broader public interest of preserving the beaches.

West of this locality are two groins that have undoubtedly complicated the picture of shore retreat in East Hampton (see chapter 4); but this does not alter the effects of the revetments.

The future will likely see additional revetments constructed on the ocean shores of the east end of Long Island despite the widely recognized fact that the shore from Montauk Point to Southampton is an eroding headland that supplies sand for beaches to the west.

Return via Ocean Avenue to Route 27.

27.8 **Traffic light. Turn left** onto Route 27.

32.2 Good view of glacial outwash plain on left and glacial moraine on right.

34.0 Traffic light in Bridgehampton.

38.9 **Turn right** at this intersection to follow Route 27 west. Route 27A goes straight ahead.

43.0 Entrance to Southampton College.

44.8 Route 27 becomes a four-lane road here.

48.7 **Take Exit 65 south** (Route 24 south).

49.4 **Turn left** at traffic light on Suffolk County Route 80 East.

50.1 **Turn right** onto Ponquogue Avenue in the center of Hampton Bays.

51.7 **Turn left** at stop sign at the end of Ponquogue Avenue onto Springville Road.

52.2 **Turn right** onto Foster Lane.

52.7 Shinnecock Coast Guard Station on left.

53.7 **Turn left** at stop sign (Ponquogue Beach).

54.7 ***Stop 5. Shinnecock Inlet parking area.*** During the storm of 1938 a washover channel developed and cut rapidly downward to produce a narrow channel that was ultimately deepened, widened, and stabilized to form Shinnecock Inlet.

A small revetment was constructed on the west side of the inlet in 1947, but it was not until 1952 that construction on the present jetties was begun. Dredging of the channel before and after construction has been carried on intermittently.

In 1938, when the inlet was opened, some sand was carried inside the inlet and deposited. Tidal currents sweeping in and out have added to that original sand mass and modified it to create the vast flood-tidal delta that exists today. At low tide, sand flats can be seen extending half the distance across Shinnecock Bay. It is all part of the flood-tidal delta, and, importantly, it represents sand that would have nourished the beaches west of the inlet had the inlet not been kept open artificially by dredging and jetty construction.

Similarly, sand being carried by the longshore currents was swept seaward to form a large asymmetric ebb-tidal delta. The edge of this delta is visible as a breaker line that swings far out to sea at the entrance to the inlet and then angles back toward the beach west of the inlet. Reasonably big waves must be approaching the shore for you to see the complete outline of this delta. Together the ebb- and flood-tidal deltas represent thousands of cubic yards of sand that should have gone to nourish beaches to the west. Moriches Inlet to the west traps sand in the same way. Between these two inlets the shoreline has retreated about five times faster than it had prior to the existence of the inlets—not a very surprising result in view of the fact that the sand supply to this portion of the south-shore barrier had been substantially reduced. As the barrier eroded, the potential for serious overwash increased, and in 1962 that fear was realized when a major northeaster hit the coast. The cry went out from local coastal dwellers to do something! The answer to this outcry is the subject of the next two stops.

Return to vehicle and proceed west down Dune Road.

57.3 To the left of the vehicle you might be able to see some blocks of asphalt resting on the marsh surface. They were carried to this position in an overwash that occurred at this point in 1972.

60.8 Solar-heated home on right.

61.6 Bridge to mainland on right.

62.8 Slow down and note the sign on the right side of the road put up by the village of Westhampton Beach. It reads:

Ordinance 39

Unauthorized trespass of any nature
on private property is subject to a
fine of $100.00. All beach property
is private.

Village Trustees

After reading this hearty welcome from the village trustees, proceed west, taking note of the absence of any parking areas for the general public and the unusual architecture of the homes. Except for the piling foundations of these struc-tures, there is little or no recognition of the fact that these buildings will need to withstand winds of 100-plus miles per hour to survive. Many seem to have been designed to fly rather than stay in place, and fly they will in the next hurricane.

64.6 West Bay Bridge.

65.7 *Stop 6*. The road narrows from four lanes to two lanes. Turn left into village parking area. Climb the wooden stairs leading to the beach and you can see the solution that was chosen for the erosion problem that was generated by maintenance of the inlets.

The groins were constructed in the late sixties in two phases. Eleven were built in the first phase, and these eleven interrupted the littoral flow of sand so that the area down-drift (to the west) suddenly began to erode. At the insistence of local interests, four more groins were built. Predictably these groins had the same effect, and there has been pressure either to continue the groin field or artificially rebuild the beaches to the west by using offshore sand. So far this pressure has been resisted.

An interesting aspect of the Westhampton Beach groin field is that it was paid for almost entirely by federal and state funds (approximately 90 percent). Despite this fact, the only public access provided is for village residents, and this is very limited. If you are from anywhere outside the village these beaches are strictly off limits for you in season

—never mind the millions you paid defending (destroying) them.

Return to vehicle and proceed west.

68.3 *Stop 7*. There is an unofficial unmarked parking lot on the right side of Dune Road at this point. It takes a sharp eye and quick reaction time to get into it, so proceed slowly to this stop. It is just beyond a two-story round home with a dome-shaped roof. Do not depend too much on the latter landmark since this home may have been destroyed by the time this guide is printed.

Remember ordinance No. 39; you are not permitted access to the beach under penalty of a $100.00 fine. If you should decide to break this law and walk to the ocean, you do so at your own risk.

You are located in the erosional shadow of the groin field visible to the east. Further west, homes are in the shelter of the east jetty of Moriches Inlet. Sand trapped by the jetty tends to keep the beach in that area stabilized, but the locality where you stand displays the dramatic effects of man's interference in a natural system. Natural dunes are nonexistent. This area is overwashed in even modest storms; the road is buried with sand; nearby homes are threatened. At one time the decks of these homes were only a few feet from the sandy surface of the barrier. This gives some idea of the amount of sand removed from this area. Now long flights of steps that can be raised and lowered to accommodate the rapidly changing surface of the sand allow access to the homes. No doubt many of these structures will be lost in the next major storm.

The last three stops of this trip illustrate how beaches and barriers can be destroyed by choosing inappropriate structural solutions. Maintenance of the inlets accelerated the problem of coastal retreat. Millions were spent on groins to slow the retreat. The groins only transferred the problem downdrift where more millions were spent for more groins. After all the expenditure the problem has not been solved, only moved further downdrift.

Clearly what is needed is a different method of dealing with our shoreline retreat. The authors of this book advocate living with barrier dynamism rather than attempting to stop it.

Trip 2. Democrat Point and Jones Beach barrier island

This field trip guide begins at the traffic circle on Fire Island just after crossing Inlet Bridge on Robert Moses Causeway. As you cross Inlet Bridge look to the left and note the location of the old Fire Island Lighthouse. Note your mileage as you enter the traffic circle at the end of the bridge. Turn west (right) following the signs to parking area no. 2.

This trip should not be taken during the peak beach season because of problems with available parking.

0.0   Traffic circle.

0.9   *Stop 1.* **Pull off to the right** of the causeway just before the road loops back toward parking area no. 2. The view north from this location is across Fire Island Inlet toward Oak Beach. The channel of the inlet has changed over the years and recently it impinged on the side where you are standing. In an attempt to stop erosion caused by the rapidly moving tidal currents at this site, slabs of concrete have been dumped to armor the shore. While this is a reasonably inexpensive technique of hardening the shore, we think you will agree that the result is less than attractive. Regardless, it has been reasonably effective in keeping the ravages of erosion from claiming the road next to which you parked your car. Continue on to parking area no. 2.

1.4   *Stop 2*. Southwest corner of parking area no. 2 at Robert Moses State Park (fig. D.1). The area where you now stand was in Fire Island Inlet in 1919 (fig. D.2). By 1936 the tip of the island extended well beyond this point. Walk west from the parking area to the end of the jetty on the east side of Fire Island Inlet. This hike takes about 35 minutes.

When you arrive at the jetty you have walked only one-quarter the distance that Fire Island has grown since 1825. This is the date the lighthouse was built, and at that time the end of the island was only 500 feet from the jetty. The walk to the jetty instills a real appreciation of the amount of material that has been deposited at this locality by longshore drift.

The jetty was built in 1940, and sand quickly began to accumulate behind it. The strips of vegetation visible in figure D.3 give testimony to the progressive infilling that occurred east of the jetty. The entire area east of the jetty was filled in only 10 years, and sand began to move around the end of the tip of the jetty and fill the inlet. This fill took the form of a series of recurved spits that were deposited west of the jetty. They are numbered in figure D.3 to show their relative age. Spit 0 is the oldest and 6 is the youngest. The amount of sand transported into the inlet is so great that it must be periodically dredged to keep the inlet safe for navigation. Figure D.4 shows the area shortly after dredging has removed the spits.

The sand removed from the spits was transferred to the beaches on the other side of the inlet. This procedure was to be repeated about every 3 years, which would have effectively created a sand bypass operation to keep the Jones Beach barrier nourished with sand. Problems with funding and environmental considerations have caused this schedule to be modified. Sand bypassing has been carried out in 1959, 1964, 1970, and 1973–1975.

What you will see on the west side of the Fire Island jetty depends on the recent activity of the U.S. Army Corps of Engineers in maintaining the inlet. If dredging has not

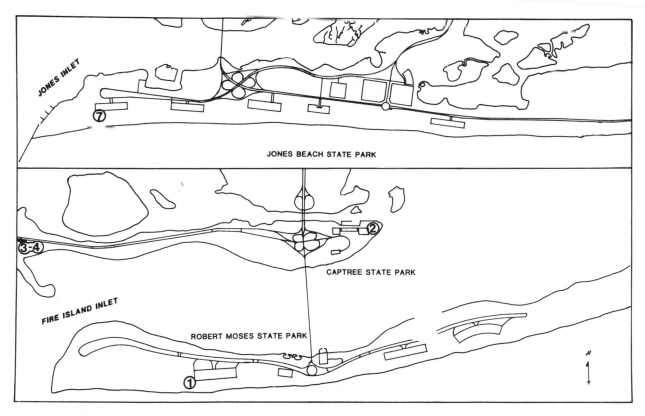

Fig. D.1. Field trip location map.

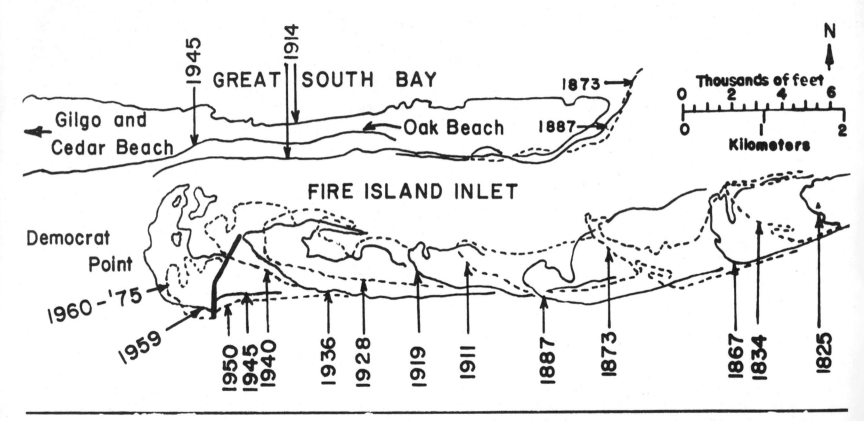

Fig. D.2. Fire Island Inlet. Source: Fred Wolff, "Trip A 5 A.M.," New York State Geological Association Guidebook, Forty-seventh Annual Meeting, 1975.

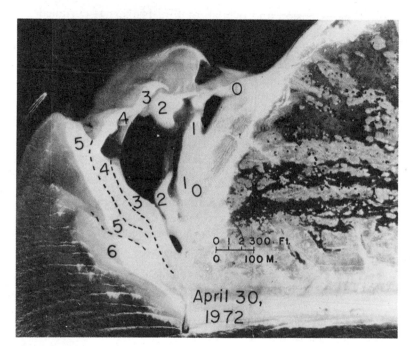

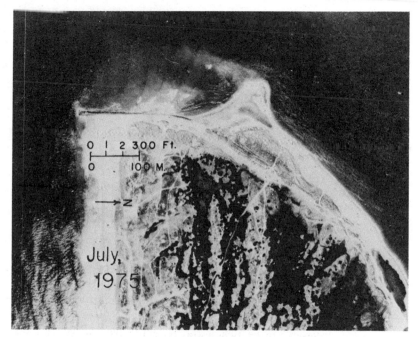

Fig. D.3. Series of spits west of the jetty on the east end of Fire Island. Source: Fred Wolff, "Trip A 5 A.M.," New York State Geological Association Guidebook, Forty-seventh Annual Meeting, 1975.

Fig. D.4. Results of dredging west of the jetty on the east end of Fire Island. Source: Fred Wolff, "Trip A 5 A.M.," New York State Geological Association Guidebook, Forty-seventh Annual Meeting, 1975.

occurred for a long period, you will be treated to a view of extensive spits curving away into the inlet. If it has just been dredged, the water line will closely border the jetty.

Return to your vehicle and proceed to the traffic circle you came around just before you entered the parking area.

2.2 **Drive around the circle** at Fire Island water tower and **cross bridge** to Jones Beach–Captree Island.

3.3 Leave bridge and follow signs to Captree Beach State Park.

3.9 *Stop 3. Captree Beach State Park*. Walk east across the dunes to the area of highest elevation. From this point you can see the end of the Jones–Captree Beach barrier island, remnants of an earlier extension of Fire Island (Sexton Island and the Fire Islands), and the position of the old and new inlets. All of this area was once exposed to the open ocean. The lighthouse on Fire Island indicates the western edge of that island and the wide inlet that was present over 140 years ago. The barrier beach west of the lighthouse and the now parallel inlet indicate the changes that have occurred since that time.

Jones–Captree Beach and its extensive backbarrier salt marsh has been dredged for the state boat channel, and large parking areas with a concession stand are provided for the public. This sets the scene for the entire barrier island. It has been developed to provide maximum recreational opportunities for the public. Preservation of the natural barrier environments is a secondary objective.

The extension and overlap of Fire Island has curtailed most erosion, though some modification by the ebb and flow of tidal currents continues. Note the bulkheads and the position of the dense scrub vegetation and trees on the dunes near the inlet. Though protected by Fire Island, it receives no sediment from that source and some erosion persists.

Captree is an important bird-nesting area and, because of its proximity to the mainland and the inlet, an important area for bathing, boating, fishing, and picnicking. Each morning a fleet of fishing vessels leaves the boat basin for points within the bay or ocean.

6.0 Leave Captree Beach and continue west on Ocean Parkway (do not turn off on Robert Moses Causeway). Turn left at the turnaround at 6 miles and proceed east on Ocean Parkway.

6.3 **Make a U-turn to the right** at Oak Beach Inn so that you are now headed back west.

7.1 *Stop 4*. Drive west until you reach a small pull-off area on the left side of the road. Pull in here and stop.

Oak Beach is a unique area for a variety of reasons. Prior to the westward extension of the end of Fire Island, this area received wave energy from the open ocean, but as Fire Island gradually overlapped this beach, all wave energy from the southeast was stopped by Fire Island. Only waves from the southwest and tidal currents affect Oak Beach. The result is that the littoral drift in the Oak Beach area is toward the east rather than the west.

The tidal currents have at times channeled their flow against the Oak Beach area, causing severe erosion. In 1959 a sand dike was built out into Fire Island Inlet to shield Oak Beach from the offending currents (area labeled "sore thumb" on figure D.1). While this effort may have been partially successful, the sore thumb acted as a dam to trap sand moving eastward from Gilgo Beach. The result is a buildup on the west side of the sore thumb and continued erosion in the Oak Beach area. The combination of channel erosion and sand starvation has driven local residents to armor their beaches with the same type of concrete slabs used at stop 1 on the other side of the inlet. The result is the ugly piles and heaps of concrete slabs and the total absence of any beach.

8.0 Return to Oak Beach Inn and **turn right** onto Ocean Parkway.

8.2 Stay in left lane of Ocean Parkway and **turn at first turnaround** so that you are again headed west on Ocean Parkway.

11.9 Cedar Beach turnaround; continue west.

15.3 Gilgo Beach; continue west on Ocean Parkway.

16.3 Hamlet of West Gilgo on right; park near western edge of this developed area after crossing Suffolk–Nassau County border.

16.7 *Stop 5*. Erosion of dunes near West Gilgo Beach. (Location is not shown on figure D.1). It is normally dangerous and unlawful to stop here on the road during the summer, and parking in the West Gilgo Beach parking lot is advisable.

This area is characterized by beach and dune erosion that has almost reached the position of the Ocean Parkway. Many areas on the adjacent barrier islands—from Fire Island to Rockaway—exhibit similar effects. The sand bypassing from Fire Island Inlet, coupled with the littoral drift, has helped to preserve this beach, but the lack of natural sand supply for the more western barrier islands (especially the Rockaways) continues to pose a serious problem. Should attempts at stabilization continue or, considering the long-range effects, should we let "nature take its course"? These are questions that have important social, economic, and political as well as environmental consequences and must be handled both on a local and a regional basis.

Leave West Gilgo and continue to Tobay Beach. As you enter the parking lot, **turn left** and follow the signs to the bird sanctuary.

17.2 *Stop 6. Tobay Beach and Bird Sanctuary*. (Location is not shown on figure D.1.) This area will illustrate the creation, development, and maturation of backbarrier, tidal salt marshes.

The "back beach" on the bay side of the barrier is fairly protected from wave action. Its intertidal zone is covered by water containing considerable suspended organic and inorganic particles. The back beach is a relatively fertile area conducive to the establishment of the salt-marsh cord grass *Spartina alterniflora.*

Patches of this tall cord grass occupy much of the upper part of the intertidal zone and act as effective sediment traps since their stems decrease the velocities of currents during the last stages of the high tide. Often the increased sedimentation in the proliferating grass may result in a shelf at the edge of the bay. This shelf is short-lived as the fringe of cord grass extends out into the intertidal zone until water depth becomes prohibitive or currents make farther outward movement difficult. Marshes are also building out into the bay, although somewhat differently, and as bay hassocks and backbeach marshes approach each other, the bay circulation is forced through narrower channels. Backbeach marshes may extend far out into the bay and perhaps even join with bay hassocks if the intermarsh currents are not strong enough to keep them separated.

The area of marsh grass at the upper end of the intertidal zone has continued to receive sediments and is in the process of building up to a tablelike surface about level with the elevation of the usual high tide. Two things now happen: first, deposition slows down, since particles can only be carried up on the marsh by the highest tides, such as full-moon (spring) and storm tides; second, salt-marsh cord grass is replaced by salt-meadow cord grass and salt grass (*Spartina patens* and *Distichlis spicata*). These secondary plants

will dominate the marsh as long as it exists, but a number of other species will invade and coexist with the dominant types mentioned above.

If the barrier beach is moving, the entire barrier island may migrate landward over the marsh. Proof of such a northerly movement on Long Island exists in the form of salt-marsh peat exposures along the ocean front. Since salt marshes cannot form in the surf area, the wave energy associated with the rising sea cuts northward, eroding the marsh deposits. Historical records indicate that extensive back-barrier-type marshes existed behind Fire Island—"hay cutting" expeditions were commercially feasible during the eighteenth and nineteenth centuries. Virtually none of these marshes exist today.

If an extensive salt marsh protects the back beach it will resist erosion. Salt-marsh peat, particularly that formed by the secondary plants *S. patens* and *D. spicata*, is a tough resiliant material. The living marsh will take an enormous beating before it disintegrates.

From the preceding discussion you can see the case for encouraging salt-marsh development on the natural areas of Jones Beach and Fire Island back beaches. It also seems that cutting mosquito-control drainage ditches into the back beach marsh is somewhat risky. One might want to encourage as complete a marsh coverage in this area as possible—even to the point of filling in larger natural waterways—to minimize weak points in this system.

Leave Tobay Beach and Bird Sanctuary and drive west on Ocean Parkway. As you drive note the patchwork of rectangular channels that slice through the salt marshes on your right. These are old dredged areas, dug to provide the fill needed for the highway on which you drive. The destruction of these tidal wetlands took place before the realization of the importance of wetlands to marine life.

This activity took place during the pave-it-rather-than-save-it era of Robert Moses. Absolutely no concession has been made to the dynamism of the natural barrier. Note the permanent nature of the structures as you approach Jones Beach.

24.4 Enter parking field 2 at West End and go to southwest corner.

Stop 7. West end of Jones Beach, parking lot 2. As with Democrat Point (stop 2), this area exhibits all the characteristic features of lateral and vertical sand accretion. The jetty constructed in the 1950s now provides a very wide beach and a wide zone of dune development. Note in particular the type of vegetation associated with these recent dunes.

The vast parking area and the four-lane highway that leads to Jones Beach are testimony to the philosophy that has guided the development of this barrier. Recreation and public access have been virtually the only considerations. Note the permanent nature of the structures erected by New York State in this area. There is little or no concession to the dynamism that characterizes natural barrier environments. It is necessary to remind yourself that you are on a barrier island. The four-lane, permanent-looking highway could as easily be located in upstate New York.

Index